CW00726628

Cover illustration

The 1.78m gabbro boulder depicting the Rhynie Man (© Aberdeenshire Archaeology Unit), which was incised into it and measures 1.03m in height, standing or walking slowly to the right, was discovered in 1978 by a farmer ploughing his field in Barflat, Rhynie. An arresting figure, the Rhynie Man was carved by the Picts of eastern Scotland, a tribal society dominated by several grades of warriors, chiefs and kings who built elaborate and well-defended settlements on hilltops or coastal promontories, and had encouraged the crafts of fine metalwork and stone carving.

The carving of the head, which represents one-quarter of the figure (as opposed to the normal one-seventh), is achieved in three basic but confident lines. Note the eyebrow, prominent nose, mouth with two large fierce-looking pointed teeth, protruding bottom lip, and the long pointed beard. Some form of head-dress or special haircut is indicated as falling onto a tunic belted at the waist. The man is wearing pointed shoes or leggings. The thin, shafted axe may be ceremonial or is perhaps a symbol of authority. The accurately carved hands show the thumbs protruding over the narrow axe shaft.

Single large figures are extremely rare in Pictish art, especially one that depicts such a solitary, almost statue-like quality. Although it is difficult to accurately date the stone carving, a date of 700–850 AD is suggested by the head-dress and preparation of the stone surface. The carving could be a memorial to a Pictish leader, a figure from a Pictish legend, or even a Christian symbol representing St. Matthew.

Source: Adapted from *The Rhynie Man*, the sixth in a series of leaflets produced by the Economic Developing and Planning Department, Grampian Regional Council, Aberdeen.

Proceedings of the
1997 Stone Weathering and
Atmospheric Pollution Network
Conference (SWAPNET '97)

Aspects of Stone Weathering, Decay and Conservation

The Robert Gordon University
15–17 May 1997

Editors

Melanie S. Jones
Roslyn Associates, Aberdeenshire

Rachael D. Wakefield
The Robert Gordon University, Aberdeen

Imperial College Press

ICP

Published by

Imperial College Press
203 Electrical Engineering Building
Imperial College
London SW7 2BT

Distributed by

World Scientific Publishing Co. Pte. Ltd.

P O Box 128, Farrer Road, Singapore 912805

USA office: Suite 1B, 1060 Main Street, River Edge, NJ 07661

UK office: 57 Shelton Street, Covent Garden, London WC2H 9HE

British Library Cataloguing-in-Publication Data
A catalogue record for this book is available from the British Library.

ASPECTS OF STONE WEATHERING, DECAY AND CONSERVATION (SWAPNET '97)

ISBN 1-86094-131-1

This book is printed on recycled paper.

Printed in Singapore by Uto-Print

Foreword

The third publication in the SWAPNET series, this volume results from the Aberdeen Network meeting held during 15–17 May 1997. Once again, it ably demonstrates the benefit of having a mechanism that puts into the public domain the early release of scientifically researched multi-disciplinary information, data and study findings.

With an international representation, the scope of the reports in the volume is both impressive and wide. This is to be warmly welcomed, as is the range of topics addressed. However, in justifying the Aberdeen venue, three scene setting papers place the Scottish dimension of the Network and development of the interest in stone conservation firmly in place. Significantly, the number of researchers studying the deterioration and other effects on a range of sandstones (which could be argued as being Scotland's predominant building stone) is expanding, and the results of their work is steadily being transmitted to the specifying world of the practitioner.

Increasing concern is being aired in official circles regarding the physical deterioration and declining state of the condition of gravestones and other carved stone subjects. It is, therefore, especially appreciated to find that this subject is beginning to receive the in-depth attention it rightly deserves. As researchers continue to be engaged in obtaining a fuller understanding of environmental matters and their impact on the surface and substrate of our buildings, it is gratifying to find that relevant conclusions are steadily emerging in this important area of work.

"Hardy annual" subjects, such as developments in our knowledge of soluble salts, lime mortar, algal growths and stone cleaning, are also dealt with in a manner that reveals the transitory nature of our understanding of previous research findings. The findings also reveal how much additional work still needs to be done.

The editors are to be congratulated for coercing, cajoling and encouraging the contributors to be so forthcoming with their material. For them to have produced such a state of the art volume so quickly after the event is to be applauded, and I am delighted that Historic Scotland has been able to be associated with its production.

Invgal Maxwell
Director, Technical Conservation, Research and Education Division, Historic Scotland
April 1998

Note: The manuscript was presented to the publisher for publication in May 1998.

Acknowledgments

The editors wish to acknowledge the following organisations/persons associated with the realisation of this volume:

Historic Scotland for providing funding towards the publication of SWAPNET '97.

The Robert Gordon University for providing a venue for the meeting and associated activities.

The Optoelectronics Research Group, The Robert Gordon University for donating funding for social activities.

The many referees who kindly gave their time to review the papers in this volume.

Chivers and Baxters Ltd for providing samples of Scottish Fare and the Tourist Board for providing information booklets.

Peter Gedge for also kindly giving his time to overcome the technical hitches experienced throughout and, finally, to all the authors of the SWAPNET '97 papers who worked hard to produce material and complete corrections of their work in time for publication.

Contents

Aspects of Stone Weathering, Decay and Conservation

Stone Weathering and Atmospheric Pollution Network '97: Aspects of Stone Weathering, Decay and Conservation.
Edited by M.S. Jones & R.D. Wakefield © 1998 Imperial College Press.

THE WEATHERING OF HASTINGS BEDS SANDSTONE GRAVESTONES IN SOUTH EAST ENGLAND

D.A. ROBINSON, R.B.G. WILLIAMS
Geography Laboratory, University of Sussex,
Brighton, BN1 9QN, England
Tel. (+44) 1273 877087
Fax. (+44) 1273 623572
E mail d.a.robinson@sussex.ac.uk

In the central Weald of south east England gravestones composed of local Hastings Beds sandstones have developed a protective silica crust. Many remain in sound condition with clear inscriptions despite over 250 years exposure to the weather. Preliminary data suggest that the western sides of the headstones weather more rapidly than the eastern sides. The weathering features displayed by the headstones differ in several important respects from those developed in the same sandstones exposed in natural outcrops or used as a building stone.

1. Introduction

Interest in the weathering of gravestones was aroused long ago by such pioneers as Geikie [1] and Goodchild [2]. Recent research has concentrated almost entirely on establishing rates of solutional loss on limestone and marble gravestone surfaces [see, for example 3-6]. Non-calcareous rocks such as sandstones, granites and slates are also frequently used as gravestones, but little systematic assessment of their weathering rates and weathering processes has been made.

This paper is concerned with the weathering of gravestones composed of local Hastings Beds sandstones from churchyards in the central Weald of south east England, an essentially rural region of villages and small market towns, which enjoys relatively low levels of atmospheric pollution. The gravestones date almost exclusively from the seventeenth to early nineteenth centuries. From 1820 onwards Permo-Triassic and Carboniferous sandstones from the Midlands and elswhere, and exotic marbles, limestones and granites, gradually supplant the use of local sandstones.

The Hastings Beds underlie much of the central Weald and comprise a complex series of sandstones, siltstones, clays and shales with subordinate ironstones and thin limestones [7]. Some of the sandstones are thin and flaggy, others are thicker and more massive. The two most massive, the Top Ashdown and Ardingly Sandstones, are highly porous, quartzose sandstones of very similar lithology, which in past centuries were widely quarried for use as building and monumental stones. They often outcrop as valley side cliffs and crags, and in recent years there have been a number of studies of the weathering of these cliffs [8-14] and of the weathering of the sandstone when used as building stone [15].

Sandstone gravestones allow the study of various weathering phenomena in a more controlled manner than is possible on natural outcrops. Firstly, it is possible to date the stones erected in the past 3-400 years from the inscriptions carved on them, or from church records, and thus gain a useful understanding of the rate at which various weathering features develop. Secondly, the gravestones provide a valuable opportunity to study the influence of aspect on weathering because they are usually erected to face east and west, in deference to the tradition that the dead must lie with their feet at the eastern end of the grave, so as to be ready to rise quickly with the first light of dawn on Judgement Day [16]. Until the nineteenth century it was customary to erect the stones with the inscriptions facing west, but then it became the norm to position the inscriptions to face the grave [17] or nearest path. Gravestones may also have varying levels of shade, insolation and leaf drip from overhanging trees, which enables the influence of these environmental factors on weathering to be studied. Finally, gravestones provide an opportunity to investigate whether the carving of inscriptions predisposes rock surfaces to certain forms of weathering.

Although gravestones provide valuable insights into rates and processes of rock weathering, they do have limitations. One is that, as sacred memorials, they should not be damaged in any way. Thus, in the present study, petrological examination has been restricted to observation and analysis under a Scanning Electron Microscope (SEM) of small fragments of stone already detached or severely loosened by weathering. A second important limitation of gravestones is that they are normally erected with the bedding planes vertical, which may predispose them to more rapid weathering than natural rock outcrops. The slimness of the stones and their high surface-to-volume ratio may also accelerate weathering damage.

2. Sandstone memorials in Sussex

In Sussex churchyards, vertical headstones embedded directly into the earth are by far the most common form of monument and for this reason they have been chosen as the main subject of this study. The most ancient headstones made of Hastings Beds sandstone are low, thick-set blocks, typically 100-200mm thick, and 400-500mm wide, that date from the seventeenth and early eighteenth centuries. The inscribed face is always smoothed, but chisel marks are sometimes visible on the rear. The inscription is sometimes placed in a recessed flat area, surrounded by a raised border. The lettering tends to be rather plain and imperfectly spaced. The tops of the stones are cut to form a single bow, a semicircle, or a triple bow.

Although a few stones rise 800mm or more above the present soil surface, many stand less than 500mm tall and the lower parts of their inscriptions lie hidden beneath the soil. Why this should be so is uncertain. In congested city churchyards repeated burials over the centuries have led to rising soil levels and partial concealment of gravestones [18], but in the relatively spacious churchyards of the Weald soil level rises are likely to have been minimal. The most probable explanation is that the stones have slowly sunk because of the yielding nature of the soil and their greater density. Earthworm and animal burrowing activity may have also helped this process.

Thick-set headstones continued to be used until late in the eighteenth century, but from the early eighteenth century, taller, thinner and more elegantly carved headstones began to be produced. These Georgian headstones are mostly 75-150mm thick, 600-950mm wide, and 850-1400mm in height. They often have extremely smooth surfaces and inscriptions that are expertly carved with bold, skilful lettering. Above the inscription it was common practice to carve a carefully crafted decoration. Favourite motifs include cherubs and angels, urns, the rising sun, hourglasses, and intertwined flowers. The tops of the stones tend to be more ornately carved than their predecessors with an abundance of concave and convex curves and scroll patterns.

These thin, elegant headstones continued in production into the second half of the nineteenth century, but from the 1830s onwards, non-local sandstone began to predominate together with Portland stone and imported marbles. By 1875 few if any gravestones were being made of Hastings Beds sandstone.

3. Weathering features

The headstones of Hastings Beds sandstone exhibit a range of features:

Surface crusting. Although the unweathered sandstone is buff coloured, stable surfaces readily discolour on exposure and become grey or black. Rough surfaces blacken less readily than smoothed surfaces, and the tops of the stones are often the least blackened. In many churchyards the most intensely blackened stones are found under yew trees, almost certainly because of some peculiarity of the chemistry of the leaf drip, which is currently under investigation by the authors.

In addition to blackening, the surfaces of the headstones develop a hardened crust and become much less permeable. Under the Scanning Electron Microscope (SEM) (Figure 1a) it is apparent that this hardening of the crust is due to the secondary deposition of silica as pore fillings and often also as a smooth coating or a surface

(a)

(b)

Figure 1: Scanning electron microscope views of a headstone dating from the 1740s in Cuckfield churchyard. 1a (above): Exterior of the blackened crust showing secondary silica deposited on and between the quartz grains.
1b (below): Section of rock immediately below the crust, showing the loosely interlocking nature of the grains.

scatter of tiny, rounded particles ranging in diameter from 1-10μm, with a mean of 4-5μm, adhering to the surface of the sand grains (typical diameter *circa* 60μm). Within the pores these silica particles congregate together to form characteristic spongiform aggregates. Some particles are almost perfectly spherical. On heavily crusted surfaces groups of sand grains up to 1-2mm across may be more or less completely enveloped in secondary silica. No oil or coal flyash particles have been observed, but algal growths and fungal hyphae are quite common.

The silica-cemented crust forms a very thin coating, at the most only a few grains in depth (150-300μm). The blackening is, if anything, even more superficial, sometimes being only about 100μm thick. Apart from the colour change, the surface crust is often difficult to detect in the field, except where with age it begins to break away from the underlying rock. Energy Dispersive X-ray analysis (EDX) indicates that in addition to silica the crust regularly contains potassium, iron and sulphur. Some samples also contain significant amounts of phosphorous, calcium and chlorine. Analysis of unweathered sandstone indicates that whilst potassium and iron are nearly always present, sulphur is generally absent, as are phosphorous, calcium and chlorine. The sand grains have relatively few adhering particles, and are poorly cemented (Figure 1b).

Both the greying and blackening of the headstones and the dissolution and redeposition of silica may be wholly or partly biological. The headstones are colonised by communities of crustose lichens, algae and fungi. Some of these are known to secrete organic acids which affect the solubility of silica and some contain fungal melanins which cause blackening of stone surfaces [19, 20]. The form of deposition of some of the silica on the surface of the sand grains suggests that it may have resulted from redeposition of silica on the surface of bacterial or fungal cells.

Granular disintegration and pitting. Grains of quartz on the surfaces of some headstones become loose and are then rubbed, washed, blown or simply fall away under gravity. This granular disintegration causes general downwearing of the surfaces of the stones, especially the tops, rounding of the edges and corners, and blurring of the inscriptions. The ancient thick-set headstones are particularly affected, and many of their names and dates are now illegible, whereas some of the taller, early Georgian stones retain near pristine inscriptions and carvings, even after 250 years.

The development of a surface crust on the headstones undoubtedly helps to reduce rates of granular disintegration. However, significant amounts of surface lowering occur even on some crusted surfaces. Figure 2a shows an eighteenth century gravestone in Worth churchyard with a blackened, crusted surface and lettering that is becoming quite faint in places. A postcard of the same stone published in about 1910 shows that the crust was already present and the inscription was much sharper (Figure 2b). Granular disintegration is making the inscription less legible despite the protection provided by the crust. Presumably, as grains become loosened and detached from the surface, fresh crust develops underneath.

Pitting develops when the surface crust is breached as a result of highly localised, rapid granular disintegration or surface flaking, possibly guided by tiny flaws or inhomogeneities (Figure 3). The pits initially have depths and diameters of only a few millimetres, but once formed readily become enlarged because they expose the softer rock behind the crust to accelerated weathering attack. The more resistant crust often protrudes as a sharp edged rim around the mouth of the pits. Pits often develop *en masse* and, as they enlarge, coalesce to form large hollowed-out areas of fresh-looking sandstone, unprotected by a surface crust.

Surface spalling and flaking. Hastings Beds headstones rarely suffer surface spalling, which following Halsey *et al.* [21] is here defined as the detachment of platy fragments more than 3mm thick. When spalling occurs it is generally because weathering has succeeded in opening up a weak bedding plane that runs parallel to the surface at very shallow depth. Flaking (the detachment of small flakes less than 3mm thick) is more common than spalling, but causes only limited damage to the stones. Thin slivers of blackened surface less than 1mm in thickness and measuring usually no more than 250mm^2 become detached, exposing the buff coloured rock beneath, which is then liable to become pitted. Multiple flaking rarely, if ever, occurs. This is in sharp contrast to the weathering behaviour of headstones made of non-local sandstones in the same churchyards which are often badly damaged by spalling and/or multiple flaking.

Cracking. Hairline microcracks, only a few millimetres deep, divide the surface crust on the main faces of some headstones into polygons, which typically are around 5-15mm in diameter. The patterning strongly recalls the crazing that develops in some ceramics. It is often difficult to see except when the sun catches the faces at an oblique angle. In some cases the cracks have become widened by weathering, and the polygons have begun to disintegrate and become detached from the underlying rock in a form of 'micro-blocky disintegration'. This form of weathering is commonest on the west faces of the headstones.

Larger-scale cracks, 10mm or more deep, sometimes form in the tops and sides of the gravestones. The two main faces of the stones are normally cut parallel to the original bedding, and the cracking tends to develop in the same direction, opening up bedding planes and partings along the tops of the stones and down the edges. In extreme cases, wedge-shaped pieces of sandstone along the edges of the stones become isolated by cracks and fall out, leaving deep slots that are then further enlarged by granular disintegration (Figure 4).

Figure 2a (upper left): Eighteenth century gravestone in Worth churchyard photographed in 1997.
Figure 2b (upper right): The same stone photographed around 1910 by an anonymous postcard publisher. Note the increased legibility of the letters, particularly in the central part of the stone.
Figure 3 (bottom left): Gravestone in Hartfield churchyard with pitting damage on the central part of its western face.
Figure 4 (bottom right): Ancient, thick-set gravestone at Mayfield showing '*delamination cracking*' which has created a slot-like opening.

Such 'delamination cracking' is particularly common on the thicker, more ancient stones. Where large cracks cut through the stones at right angles to the main faces, 'blocky disintegration' or 'splitting' may occur, but this is much less common than delamination cracking.

Some larger-scale cracks form irregular and incomplete networks, but never the well organised polygonal patterns that occur on some natural sandstone outcrops [22]. However, faint polygons resembling the outcrop patterns have formed on the sandstone ledgers of Georgian box tombs at Balcombe.

Hairline microcracking on the headstones may result from stresses set up when extra silica deposited in the crust by evaporating pore water dries and shrinks, or from moisture movements or temperature changes. Its occurrence on the tops and west faces of headstones that receive the greatest sunshine strongly suggests that insolation is important. The larger-scale cracking is more difficult to explain, but may be a delayed stress release phenomenon acting on flaws within the sandstone.

4. Weathering processes and time

The amount of weathering varies considerably between individual headstones and between churchyards. There is no obvious relationship between the age of a headstone and the extent of weathering. All the more ancient thick-set headstones have undergone some weathering, though there is no way of assessing the actual amounts. The stones are extensively crusted, and very little surface pitting or breaching of the crust has taken place. The inscriptions are often still visible, though many are illegible or can be read only with difficulty due to continuing surface loss caused by granular disintegration. Although the most conspicuous evidence of weathering are the networks of cracks that are found on the tops and edges of certain stones, granular disintegration is probably the most important weathering process which leads to a gradual downwearing of the surfaces and rounding of the edges.

Some tall Georgian headstones are impressively preserved, with near pristine surfaces and clear, sharp inscriptions. Others, however, show obvious weathering damage. The inscriptions have become faint, and sometimes unreadable, the ornately carved decorations blurred and blunted through granular disintegration. The main faces of the more badly affected headstones are disfigured by a combination of granular disintegration and pitting, particularly in the centres, away from the soil surface and the edges of the stones (Figure 3). It is rare for weathering attack to progress further, but a few stones have reached such an advanced state of decay that only their tops and base retain blackened crust while the rest of the surface is unstable, crumbling and deeply hollowed out (Figure 5a). Sideways views reveal that these severely weathered stones have undergone most thinning at or just above their mid-line (Figure 5b).

It is not entirely certain why the centres of the faces suffer greatest damage. One might have expected that rising damp and salt attack would concentrate weathering close to the soil level causing *'basal rotting'*. Visible salt deposits are uncommon but where they occur, as at Worth, efflorescences coat the base of the stones, especially on the eastern sides. Analysis of the deposits at Worth shows that the dominant salt is hydrated calcium sulphate (gypsum) although small quantities of sodium salts are also present. The rock surfaces beneath the deposits appear quite sound, so it is doubtful in this case that the salts cause damage.

The reason why the centres of the faces of the gravestones weather most may lie in the fact that they blacken particularly readily or in some peculiarity of their moisture regime. The base of the gravestones may retain moisture for such long periods as to inhibit weathering. The centres of the faces may experience a greater number of wetting and drying cycles, yet still remain damp for long periods, aided perhaps by a better-developed and more impervious crust. The exposed tops and edges of headstones tend to dry out more quickly after rain than the rest of the stones, allowing them to resist weathering. In damp conditions, it is easy to observe that it is the central parts of the stones that are wettest for longest, in part, probably as a result of moisture wicking up into the stone from the soil beneath. Experimental investigations of the wick effect have demonstrated that on small specimens of rock, maximum weathering occurs around the limit of capillary rise [23].

It is a remarkable fact that all the most severely weathered headstones are located beneath yews and other conifers (Figure 5a and b) or close to south and east facing church walls. The first location suggests that shade and/or the chemistry of leaf drip is a major determinant of weathering, the second that the soil moisture regime may be important, dry walls tending to draw water through the soil towards their evaporating surfaces.

The damage caused by yews and other conifers is not consistent. Next to a severely weathered headstone and under the same yew or adjacent yews there are often other headstones that are virtually pristine, and in some churchyards none of the headstones under yews shows more than average amounts of damage. Evidently, whatever factor causes severe weathering of headstones under conifers operates in combination with other factors that can completely nullify it.

5. Aspect and weathering amounts

The influence of aspect on weathering intensity has been assessed here by selecting sample graveyards and comparing the volume of surface loss on the eastern and western faces of the ten most weathered headstones that are located in the open, away from trees and the church walls. The volume of loss could in theory be measured using a Talysurf plotter [24, 25] or laser micro-mapper [26], but this would be highly time-consuming and create spurious accuracy given the difficulties of determining the precise original dimensions of the headstones. The method adopted in the present study has been to measure maximum depths of weathering (or, when this by itself would yield misleading results, then also the areas) in order to determine which face of each headstone has suffered the greater volume of surface loss. A steel rule is rested on the parts of each face that most nearly approximate to the original surface and then the depth from the rule into the deepest pit or hollow on the surface is measured using an engineer's gauge that reads to 0.1mm. Care is taken to avoid obvious flaws in the stone, carved inscriptions and ornamentation. One drawback with the method is that if the entire face has been reduced by surface granular disintegration the measurement obtained will underestimate the true maximum depth. However, this is not considered a major problem, because even the most weathered Hastings Beds gravestones retain some fragments of largely undamaged surface.

Preliminary results from six churchyards that contain representative samples of weathered headstones are shown in Table 1. In four of the six churchyards western faces are markedly more weathered than eastern (though in one, Lingfield, there is a tie). Only at Worth are eastern faces more weathered than western. In the majority of cases it is the western face that carries the inscriptions, but the deepest pits or hollows are not always located within the area of the inscriptions, and there is no reason to suppose that the carving of the inscriptions has promoted subsequent weathering damage. At Ticehurst, a sample of 5 headstones with their inscriptions facing east still exhibit maximum weathering on the west. The data suggest that weathering rates are on average greater on the western faces than on eastern, but there are evidently some differences between churchyards. Very different results have been obtained for the non-local sandstone; these are generally much more deeply weathered on their eastern than western sides. Weathering is also greater on the eastern sides of Sussex church towers [15].

The tendency of the Hastings Beds gravestones to be most weathered on their western sides could be due to various factors. The western sides face the prevailing wind and are more frequently rain-soaked; the sheltered eastern sides tend to be much drier.

(a) (b)

Figure 5: Back (a), and side (b) views of a severely weathered gravestone under an old redwood (*Sequoiadendron giganteum*) in Cuckfield churchyard. Note the preservation of dark crust at the top and to a lesser extent the base of the stone, and typical "high-waisted" profile.

Churchyard	Face with inscription		Dates of headstones	Face with maximum weathering			Mean maximum depth (mm)	
	West	East		West	East	Equal	West	East
Worth	9	1	1802-1857	1	9	0	1.7	4.2
Lindfield	5	0	176?-1848	3	2	0	6.1	4.1
Hartfield	10	0	1774-1858	8	2	0	3.5	1.4
Mayfield	10	0	1737-1847	8	2	0	2.6	1.2
Lingfield	10	0	1759-1838	5	5	0	2.5	2.0
Ticehurst	10	0	17??-1843	6	1	3	3.6	3.3
Ticehurst	0	5	1760-1817	5	0	0	2.4	1.1
Total/mean/ range	54	6	1737-1858	36	21	3	3.0	2.5

Table 1. Maximum depths of weathering on Hastings Beds headstones

The eastern sides face the morning sun and this may result in their starting to dry first after overnight rain. On the other hand, the western sides are sun-lit in the second half of the day when air temperatures are highest, and this may cause them to become drier than the eastern sides despite their delayed start. These factors clearly require further investigation.

6. Comparison with weathering on natural outcrops

The formation of silica-cemented crust on the sandstone gravestones (both in the letters and other surfaces) is of particular interest. The crust is almost as well developed as it is on natural outcrops of the sandstone in the Weald, despite earlier remarks to the contrary [13]. The speed of crust formation is surprising, given the relative insolubility of silica and the small volume of rock represented by the gravestones. Rainwater entering the outcrops could be expected to travel considerable distances through the rock perhaps dissolving silica before returning to the surface and depositing the silica through evaporation. The gravestones offer only a limited volume of rock within which rainwater can circulate, and yet they have developed significantly hardened surfaces within 150 years in some cases. This suggests that biogenic processes may be particularly important in the formation of the crust.

Despite the similarities provided by crust formation, the gravestones differ from the natural outcrops in the relative abundance of some weathering features. The most notable difference is the absence on the gravestones of honeycombing or alveolar weathering which is widespread on the natural outcrops and also occurs on sandstone blocks used to build local churches [15]. In part this may be due to differences in the age of the rock surfaces. Some of the honeycombing on the outcrops has crusted over and could well be many centuries old. At Eridge Rocks, for example, a large area of well-developed honeycombing has not changed in any discernible way in the hundred years that have passed since it was photographed in 1897. However, honeycombing on some Sussex churches built of Hastings Beds sandstone has developed within a hundred years.

It might seem that the absence of honeycombing on the gravestones could be due either to their thinness or the vertical orientation of the bedding planes. The trouble with both explanations, however, is that honeycombing can sometimes be found in Wealden churchyards on gravestones of non-local sandstone of equivalent or younger age that are not only thin but have vertical bedding planes. It is therefore likely that the stonemason selected particularly tough layers of Hastings Beds sandstone to make gravestones, and these layers are relatively unsusceptible to honeycomb weathering.

Surface spalling occurs more frequently on the natural outcrops than on the gravestones, particularly where the rock has more than average fissility. Again, the difference may simply be due to the stonemason selecting the more durable freestone for gravestones. Alternatively, it may reflect the relative young age of the gravestones. Spalling rarely afflicts Sussex sandstone churches built in the last 200 years and only becomes a noticeable feature on churches that are at least 500 years old [15].

The small-scale polygonal cracking or crazing that affects the faces of some gravestones has not yet been found on the natural outcrops, though it has been observed in northern France in setts cut from Fontainebleau Sandstone [22]. The larger-scale crack patterns that develop on the tops of the headstones quite closely resemble the polygonal cracking that occurs on some outcrops [10, 22, 27], although on headstones they are more irregular and incomplete. It is notable that the headstones exhibit silica crusting which Williams and Robinson [27] identified as a common attribute of those natural exposures of rock that exhibit polygonal cracking.

The cessation of weathering and restabilisation of surfaces that have suffered appreciable weathering and erosion in past times is a puzzling feature common to natural exposures, building stones and headstones alike. Why it occurs is uncertain, but it is probably due to changes in the local environment to which the surfaces are exposed. Especially important may be the growth or death and removal of trees or other vegetation which could greatly alter insolation or shade levels, and in some cases significantly change the chemistry of rainwater falling onto the rock surfaces.

7. Conclusions

Headstones made of local sandstones show remarkable resistance to weathering. Inscriptions dating back to 1634 are still capable of being read without much difficulty, and the carving on some stones from the second half of the eighteenth century is still near pristine. This is particularly surprising given the poor reputation of the local sandstone as a building stone, although it accords with the way the sandstone forms impressive cliffs along many valley sides in the Weald. The ability of the gravestones to form a silica enriched crust appears to be a key factor in their survival. Even the youngest stones, which date back to around 1875, show significant crust formation, so the process is evidently quite rapid. The weathering forms displayed by the headstones are similar to those exhibited by the natural outcrops and blocks of sandstone used for building, except that honeycomb weathering is absent, spalling is less common and the crack patterns show important differences. As on the natural outcrops, eroding surfaces can sometimes stabilise and crust over. Greatest damage occurs beneath yew trees and other conifers, the reasons for which require further investigation.

8. References

1. Geikie A., Rockweathering, as illustrated in Edinburgh churchyards. *Proceedings of the Royal Society of Edinburgh* **10** (1880), 518-532.
2. Goodchild J.G., Notes on some observed rates of weathering of limestones. *Geological Magazine* **27** (1890), 463-466.
3. Rahn T., The weathering of tombstones and its relation to the topography of New England. *Journal of Geological Education* **19** (1969), 112-118.
4. Meierding T.C., Marble tombstones weathering rates: a transect of the United States. *Physical Geography* **2** (1981), 1-18.
5. Klein M., Weathering rates of limestone tombstones measured in Haifa, Israel. *Zeitschrift fur Geomorphologie* **N.F.28** (1984), 105-111.
6. Cooke R.U. Inkpen R.J. and Wiggs G.F.S., Using gravestones to assess changing rates of weathering in the United Kingdom. *Earth Surface Processes and Landforms* **20** (1995), 531-46.
7. Gallois R.W., *The Wealden District*. British Regional Geology (HMSO, London, 1965).
8. Bird E.C.F., *Tor-like sandrock features in the central Weald* (Abstract) (20th International Geography Congress, London 1964), 1156.
9. Piper D.J.W., Pleistocene superficial deposits, Balcombe area, central Weald. *Geological Magazine* **108** (1971), 517-523.
10. Robinson D.A. and Williams R.B.G., Aspects of the geomorphology of the sandstone cliffs of the central Weald. *Proceedings of the Geologists Association* **87** (1976), 93-100.
11. Robinson D.A. and Williams R.B.G., Sandstone cliffs on the High Weald landscape. *Geographical Magazine* **53** (1981), 587-92.
12. Robinson D.A. and Williams R.B.G., *Classic Landforms of the Weald.* (Landform Guides No 4, Geographical Association, Sheffield, 1984).
13. Robinson D.A. and Williams R.B.G., Surface crusting of sandstones in southern England and northern France. In *International Geomorphology 1986, Part 2*, ed. Gardner V. (Wiley and Sons, Chichester, 1987), 623-635.
14. Pentecost A., The weathering rates of some sandstone cliffs, central Weald, England. *Earth Surface Process and Landforms* **16** (1991), 83-91.

15. Robinson D.A. and Williams R.B.G., An analysis of the weathering of Wealden sandstone churches. In *Processes of urban stone decay*, ed. Smith B.J. and Warke P.A. (Donhead, London, 1996), 133-149.
16. Lindley K., *Of graves and epitaphs* (Hutchinson, London, 1965).
17. Beevers D. Marks R. and Roles J., *Sussex churches and chapels*. (Royal Pavilion Art Gallery and Museum, Brighton, 1989).
18. Burgess F., *English churchyard memorials*. (Lutterworth, London, 1963).
19. May E. Lewis F.J. Pereira S. Tayler S. Seaward M.R.D. and Allsopp D., Microbial deterioration of building stone - a review. *Biodeterioration Abstracts*, **7** (1993), 109-123.
20. Saiz-Jiminez C. Ortega-Calvo J.J. and de Leeuw J.W., The chemical structure of fungal melanins and their possible contribution to black stains in stone monuments. *Science of the Total Environment* **167** (1995), 305-314.
21. Halsey D.P. Dews S.J. Mitchell D.J. and Harris F.C., The black soiling of sandstone buildings in the West Midlands, England: regional variations and decay mechanisms. In *Processes of urban stone decay*, ed. Smith B.J. and Warke P.A. (Donhead, London, 1996), 53-65.
22. Robinson D.A. and Williams R.B.G., Polygonal weathering of sandstone at Fontainebleau, France. *Zeitschrift fur Geomorphologie* **N.F.33** (1989), 59-72.
23. Goudie A.S., Laboratory simulation of 'the wick effect' in salt weathering of rock. *Earth Surface Processes and Landforms* **11** (1986), 275-285.
24. Young M. and Urquhart D., Abrasive cleaning of sandstone buildings and monuments: an experimental investigation. In *Stone cleaning and the nature, soiling and decay mechanisms of stone*, ed. Webster, R.G.M. (Donhead, Aberdeen, 1992), 128-140.
25. Moses C.A., Methods for investigating stone decay mechanisms in polluted and 'clean' environments. In *Processes of urban stone decay*, ed. Smith B.J. and Warke P.A. (Donhead, London, 1996), 212-227
26. Swantesson J.O.H., Micro-mapping as a tool for the study of weathered rock surfaces. In *Rock weathering and Landform Evolution*, ed. Robinson D.A. and Williams R.B.G. (J. Wiley, Chichester, 1994), 209-222.
27. Williams R.B.G. and Robinson D.A., Origin and distribution of polygonal cracking of rock surfaces. *Geografiska Annaler* **71A** (1989), 145-159.

8. Acknowledgements

The authors thank Mrs. N. Thomson for carrying out the chemical analyses and Mr. D.P. Randall for assistance with the SEM and EDX. Valuable comments were received from an anonymous referee.

Stone Weathering and Atmospheric Pollution Network '97: Aspects of Stone Weathering, Decay and Conservation.
Edited by M.S. Jones & R.D. Wakefield © 1998 Imperial College Press.

GRAVESTONES: PROBLEMS AND POTENTIALS AS INDICATORS OF HISTORIC CHANGES IN WEATHERING

R. J. INKPEN
Department of Geography, University of Portsmouth, Buckingham Building
Lion Terrace, Portsmouth, Hampshire, England, PO1 3HE
Tel 01705 842467 Fax 01705 842512
E mail inkpenr@geog.port.ac.uk,

Gravestones have been used to assess historic weathering rates by a number of authors. Some of the problems with measurement of gravestones that could influence the derived weathering rates are outlined. In particular the assumptions underlying the lead lettering index are explored and the importance of micro-weathering environments illustrated. Despite these problems the technique still has potential to permit comparison of historic weathering losses and rates between locations. Using a range of methods, the presence of an industrial/urban/rural gradient in weathering rates is explored. The data suggest that although such gradients may exist, the use of land-use to define these past weathering environments needs careful consideration.

1. Introduction

Gravestones have been used in previous studies to determine historic weathering rates and their variability. Individual studies [1-3] set out to compare the impact on weathering loss of different atmospheric pollution histories. Attewell and Taylor for example, quantify the spatial difference in weathering rates over time between rural, urban and industrial areas of the Durham and Newcastle region [2]. Cooke *et al.* compared weathering losses between three locations with differing weathering and environmental histories to assess if difference in the relationship between age and weathering could be identified between locations [3]. They found that the two sites, classed as urban, Portsmouth and Wolverhampton, although differing in mean weathering rates had statistically similar linear relationships between weathering loss and age. Swansea, the location classed as industrial, however had a much steeper linear relationship between weathering loss and age as well as having a higher mean weathering rate. Despite higher rainfall in Swansea than in the other two sites over the period of interest, this implied that weathering loss over time was influenced by the pollution history of a location.

Cooke suggested that there were two trends in weathering rate that would be expected given the history of polluted atmospheres [4]. Temporally there would be a rise in sulphur dioxide and consequent increase in weathering rates and loss as industrialisation increased. More recently in Western countries there might be a slight decrease in levels of atmospheric pollution as clean air legislation take affect. This would result in a decrease in weathering rates and loss. Secondly, there would be a spatial trend, with atmospheric pollution and weathering rates being highest in urban areas. Identification and quantification of these trends requires a source of data that can extend information about weathering behaviour of stone back in time and yet provides samples over a wide area. Gravestones, and in particular Carrara marble gravestones, potentially provide such a data source. The work presented here outlines a method for quantifying weathering loss from gravestones (the lead lettering index) and explores its limitations both in terms of its internal consistency and for comparative studies, an essential element in assessing the trends identified by Cooke [3].

2. Measurement methods

Gravestones, in particular those made of Carrara marble, have been used to identify and quantify weathering rates between different areas of the country. The practice of using lead lettering in some of these gravestones has been useful in providing a baseline from which surface loss can be measured. Lead lettering was inserted into the gravestones in grooves originally cut for the lettering. The marble was polished flush to the lead lettering and so the lettering provided an indication of the original level of the gravestone. It is assumed that the lead lettering is relatively stable over time, whilst the marble is relatively unstable and subject to loss from weathering processes. The amount of marble lost is assumed to be dependent upon both the duration over which the surface is exposed to weathering processes and the intensity of those weathering processes. During periods of high atmospheric pollution it is assumed that a higher rate of atmospheric deposition occurs resulting in higher losses of marble from the gravestone. Cooke *et al.* [3] used this technique for assessing differences in weathering rates between three sites, Swansea, Portsmouth and Wolverhampton. In this study three different letters on the gravestone surface were measured. For each letter four point measurements were taken around each letter, one at the top, one at the base and one on each side of the lettering. Using '1's and 'E's ensured that the same letter on each gravestone was measured and they were usually in the same location on the gravestone, i.e., '1' in the date of birth and death and E' in the word 'DIED'. To ensure similarity of exposure conditions several conditions had to be met for any gravestone selected. Any gravestone showing evidence of relocation was not included in the study as the weathering environment it had been exposed to was likely to have varied. For similar reasons tilted or horizontal gravestones were not included in the study. In addition, gravestones under trees were excluded from study as they were prone to the effects of leaf drip which delivered rainfall of potentially higher pollutant concentration and intensity. Lastly, gravestones with lead lettering that was curling and peeling from the face were excluded from the study. Measurement of surface change from these letters would produce erroneously high amounts of loss.

2.1. Potential problems associated with lead lettering index

Despite restrictions on gravestone selection there are still a number of potential problems with using the Cooke *et al.* technique to assess past weathering. Firstly, there is the assumption of the lead lettering being flush with the marble surface at the start of weathering. Although personal discussions with monumental masons suggest that this was and is the practice, there is no guarantee that the level of polishing was consistent either within the same graveyard, between time periods or across the country. The potential impact of even slightly different starting conditions on subsequent weathering losses is unclear.

Secondly, the lettering itself can create micro-weathering environments that could produce different weathering losses on the gravestone surface. Figure 1 is a digital terrain model of the surface between and around two letters, N and O, on a gravestone. Using close range photogrammetry it is possible, provided there are sufficient control points, to reconstruct the surface to an accuracy of better than 35 microns [5-7]. Across the relatively small area reconstructed there is great variation in surface heights and, importantly, a great deal of variation about the lettering itself. This does tend to suggest that micro-scale weathering environments are present, even at this scale, and produce effects of sufficient magnitude that could influence the losses measured.

Thirdly, there is the problem of some gravestones having non-parallel faces. This means that a slight slope might be present on some gravestones resulting in differences in runoff and contact times and, potentially, therefore in weathering losses. Lastly, there is an assumption that a continuous measurement of surface loss can be made. The curling and detachment of lettering partly occurs because surface loss occurs. As the surface retreats, the marble into which the pegs of the lettering are fixed is lost. Over time, the lettering loosens and becomes detached. This could produce erroneously high weathering losses as the lettering rises from the surface. The presence of such bowing or curling could imply that there is a limit to the depth of loss lead lettering can indicate. As the lettering becomes detached and curls this represents the limit of acceptable measurement and so is controlled partially by the original cut depth of marble into which the lettering was placed. At Swansea [3] this type of behaviour has not occurred. The relatively high rate of loss of marble has instead left the lead lettering on small pedestals proud of the retreating surface. This means that the original peg holes in the marble have not been affected by surface loss, as the whole area covered by the lead lettering has been protected from surface loss. Only in these sets of weathering conditions will lead lettering provide a measure of surface loss beyond the limit of the original groove depths.

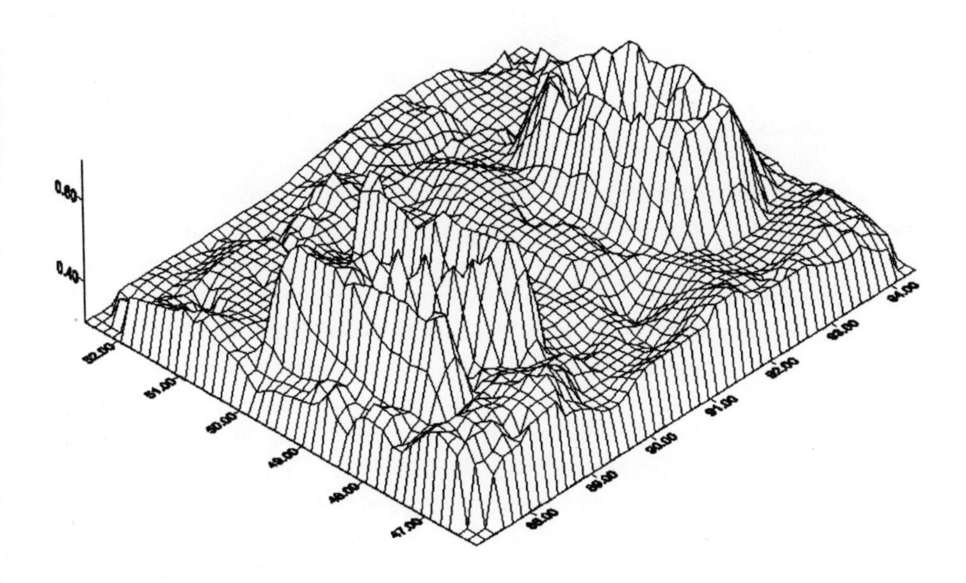

Figure 1. Digital terrain model of the surface between and around two letters, N and O, on a gravestone. X and Y axis in millimetres, Z axis in microns, Z values refer to height above a reference plane.

3. Experimental design to assess weathering trends

The spatial and temporal trends suggested above can be assessed by designing a sampling strategy that permits the comparison of weathering data from a range of sites. The work presented here combines data from other studies the author has been involved in [3, and unpublished work] to continue the assessment of the impact upon weathering loss and rates of variations in pollution history, and in other environmental variables across the United Kingdom. Nine sites are used with data being collected at different times from each site (Table 1). Using these sites two spatial patterns can be identified and quantified and some indications of a third obtained. Firstly, the impact of changing historical pollution levels can be assessed by comparing industrial, urban and rural sites. Secondly, the impact of marine influences such as sea salts can be assessed by comparing coastal and inland sites. Some limited assessment of inter-site variability can also be carried out between urban and rural sites in different parts of Southern Britain.

4. Data Analysis

There are a number of means of assessing the degree of similarity or difference between locations. Firstly, the mean weathering rates for each site can be calculated on the basis of data from individual gravestones. This involves determining the mean depth of loss from the twelve measurements from each gravestone and then dividing the result by the age of the gravestone. Multiplying the previous calculation by 1000 converts the results from mm to μm. Table 2 presents the results of these calculations.

An initial survey of the data suggests that there is a difference in weathering rates between the industrial locations, Swansea and the Jewellery Quarter, Birmingham, and the other urban and rural locations. There is a less distinct difference, if at all, between the various urban and rural locations, except for the very low weathering rate for the inland rural location.

Table 1. Site key and survey date and sample size.

Site Location	Site Key	Land-use Type	Date of Survey	Sample Size
Wolverhampton	Wo	Urban Inland	1992	17
Portsmouth	Po	Urban Coastal	1997	75
Swansea	Sw	Industrial Coastal	1992	34
Jewellery Quarter, central Birmingham	JQ	Industrial/ Urban Inland	1996	17
Clacton-on-Sea	Cl	Urban Coastal	1996	100
Oxford	Ox	Urban Inland	1996	100
Rural sites near Clacton	CR	Rural Coastal	1996	100
Rural sites near Oxford	IR	Rural Inland	1996	100
Lodge Hill, Birmingham	LH	Urban Inland	1996	59

Table 2. Mean weathering loss and weathering rates for each location (μm and μm/year).

Location	Mean Loss μm	Std. Dev.	Mean Weathering Rate μm/year	Std. Dev.
Wolverhampton	510	154	6.820	1.472
Portsmouth	719	140	8.172	1.274
Swansea	2105	1237	26.785	10.783
Jewellery Quarter	1912	793	20.072	6.715
Clacton	444	162	6.360	1.347
Oxford	634	306	8.206	2.633
Coastal Rural	461	171	6.042	1.217
Inland Rural	251	146	3.175	1.419
Lodge Hill	648	212	8.341	2.181

Table 3 outlines the results of a Mann-Whitney U comparison of the data between pairs of sites. Most locations seem to have statistically significantly different mean weathering rates relative to other locations. Only a few similarities are present between sites, most notably urban sites, whether coastal or inland have a similar weathering rate to at least one other urban site. Weathering rates at the coastal rural site, unlike rates at their inland counterparts, are of a similar level to those found in some urban areas, notably the closest urban site, Clacton, and the suburban site of Wolverhampton. This suggests that although the industrial, urban, rural weathering pattern is present it is not as clear-cut as might be expected. Specifically, industrial locations have significantly greater weathering rates than the other sites. Urban and rural sites do have statistically different weathering rates, but the relationship is more context dependent. The inland urban and rural pairing in Oxfordshire do have a statistically significant difference with Oxford having the higher weathering rate. Similarly, urban sites tend to have similar weathering rates, but there are significant differences between some locations. This implies that a simple three-fold division based on land-use is not necessarily reflected in weathering rates.

Table 3. Mann-Whitney U test comparison of weathering rates between sites

Location	Urban	Rural	Industrial
Wo	Cl	CR	None
Po	Ox, LH	None	None
Sw	None	None	None
JQ	None	None	None
Cl	Wo	CR	None
Ox	Po, LH	None	None
CR	Wo, Cl	None	None
IR	None	None	None
LH	Po, Ox	None	None

Pairs with statistically similar weathering rates ($\alpha = 0.05$)

A comparison of mean weathering rates only assesses the mean and the spread of data about that mean, it tells us very little about the similarity or difference between locations in terms of the relationship between age and weathering loss. Table 4 outlines the linear regression equations for all the sites. None of the linear regression equations intersect the y-axis at zero. This implies that at least in the initial stages of weathering there is not a linear relationship between weathering loss and age. Such a non-linear relationship in the early stages of weathering for marble gravestones was suggested in Israel [8]. Klein found that the initial surface polish retarded weathering in the initial years of exposure producing a relatively low amount of surface loss. Once the surface polish had been removed, loss would increase and eventually reach a linear relationship with age. Relationships with negative intercepts would imply that this suggestion might operate. A positive intercept, such as found at Portsmouth, could be explained by the same mechanism. Initially low loss might be followed by high loss once the surface polish is removed. Exposure of fresh surfaces could result in rapid weathering and removal of material before an equilibrium between process and material response is established. The initial high losses may be sufficient to alter the linear regression equation to a positive intercept value.

Table 4. Regression equations for each location for weathering loss/age relationship (in μm).

Location	Constant	Gradient	r^2 value
Wolverhampton	133	4.93	55%
Portsmouth	206	5.78	41%
Swansea	-1250	46.6	77%
Jewellery Quarter	48	19.7	38%
Clacton	-321	11.3	81%
Oxford	-553	16.1	70%
Coastal Rural	-209	9.03	82%
Inland Rural	-169	5.6	69%
Lodge Hill, Urban	39	7.81	42%

Using analysis of covariance it is possible to compare the similarity or difference between the regression lines derived for depth of loss and age for each location [9]. Analysis of covariance compares the regression line for data from each location to that derived for one anchor location. By varying the anchor location, the significance of the relationships between all locations can be determined (Tables 5a and 5b). Before analysis of covariance is carried out all the data are combined into one data set and the average age calculated. The average age is then subtracted from the age of each gravestone for all locations. In this manner, the regression line for each site passes through the y-axis at x=0, or the average age of a gravestone for the whole data set. This aids interpretation as the intercept value can be related to a gravestone of average age at each site.

Table 5a outlines the relationship between different anchor locations and other locations for the constant, i.e., the intercept, in the regression equation. Comparing the data, an average age gravestone at Swansea has a significantly greater loss than an average aged gravestone at any other location. A similar conclusion can be reached for an average aged gravestone from the Jewellery Quarter, Birmingham. Elsewhere, the intercept table confirms the differences and similarities identified using the Mann-Whitney U test. Table 5b outlines the differences in the significance of the gradient of the regression line between locations, this represents the rate of change of loss between locations. The significance of Swansea is again apparent since this location has a higher rate of loss than any other location in the study. The Jewellery Quarter, Birmingham has a significantly greater rate of loss than any rural or urban site, other than Oxford. A complex pattern of significance is found amongst the other urban locations. Wolverhampton, Portsmouth, Lodge Hill, Birmingham, and the two rural sites all have rates of loss similar to each other. Clacton-on-Sea has a similar rate of loss to its rural counterpart and to Lodge Hill, Birmingham. This data suggests that although they may differ in the amount of surface loss for the average aged gravestone, most urban sites have rates of loss similar to each other and, in some cases, rates that are not significantly different from rural locations. In addition, formerly industrialised locations have higher rates of loss than urban sites. Oxford is an interesting exception to this pattern. It has a significantly higher rate of loss than the other urban sites and appears to have more in common with the Jewellery Quarter than with the urban sites. This may reflect particular aspects of its past weathering environment, for example, the presence of light industry nearby in car manufacturing and the relatively built-up nature of the surrounding area compared to the contemporary relatively suburban location of the other urban graveyards. It also highlights the importance of not taking the data from studies of gravestones in isolation. In order to develop an appropriate interpretation it is necessary to have some background knowledge of the local history of the location. This knowledge can then be used to assess how appropriate the classification of each location is with respect to its past. For Swansea for example, the site next to the graveyard had been occupied by a copper smelter until the early part of this century.

Table 5a. Analysis of covariance of data from nine sites: Intercepts.

Anchor Location	Urban	Rural	Industrial
Wo	Cl	CR	None
Po	Ox, LH	None	None
SW	None	None	None
JQ	None	None	None
Cl	Wo	CR	None
Ox	Po, LH	None	None
CR	Wo, Cl	None	None
IR	None	None	None
LH	Po, Ox	None	None

Locations with statistically similar weathering loss intercepts ($\alpha = 0.05$).

Table 5b. Analysis of covariance of data from nine sites: Gradients.

Anchor Location	Urban	Rural	Industrial
Wo	Po, LH	CR, IR	None
Po	Wo, LH	CR, IR	None
Sw	None	None	None
JQ	Ox	None	None
Cl	LH	CR	None
Ox	None	None	JQ
CR	Wo, Po, Cl, LH	None	None
IR	Wo, Po, LH	None	None
LH	Wo, Po, Cl	CR, IR	None

Locations with statistically similar gradients of loss ($\alpha = 0.05$).

In the Jewellery Quarter of Birmingham there was a high concentration of artisan workshops producing heavy metal and smoke pollution until the 1950s. Stone weathering data losses from these locations were interpreted in the light of this local information. Similarly, losses at Lodge Hill, Birmingham were not expected to be great as the graveyard was a large and relatively open, away from current major industrial sources in an area of suburban housing dating from between the turn of the century to the 1960s. The importance of this local knowledge lies in its use for assessing the appropriateness of the classifications applied for comparison. Wolverhampton, Portsmouth, Clacton-on-Sea and Lodge Hill, Birmingham are all classified as urban. In reality their locations within the urban area are suggestive of a suburban history. In this case it is not a surprise that they are similar. The Jewellery Quarter could also be described as urban today, it is only a knowledge of its past that permits a reclassification as *small scale* industrial during the time period covered by the gravestones. This suggests that as well as collecting data on the weathering of gravestones, an understanding of a locations local history is vital in data interpretation.

5. Conclusion

This work assesses the similarities and differences between weathering histories derived from gravestone surface loss data from nine locations in Southern Britain. All data were collected using the same technique, the lead lettering' index, performed on gravestones of the same lithology, Carrara marble. Some problems and assumptions made when collecting and analysing this type of data have been outlined. There appears to be a limit at most sites to the amount of loss that the lead lettering technique can record. Even where losses can be measured, investigators should be cautious in interpreting micro-scale variations in loss and their potential impact upon measured weathering rates. When comparing the mean weathering rates and linear regression equations between sites some consistent trends emerge. The industrial sites are statistically different from all other sites having higher weathering rates and losses. The urban/rural distinction between weathering rates and loss appear to be context specific, being observable between Oxford and its rural counterparts, but not between Clacton-on-sea and its rural counterparts. There also appears to be differences between urban sites in their weathering behaviour. This could imply that more care is required in defining and assessing what is meant by the different land-use categories and their stability over time. Alternatively it could point to the need for a more contextual and site-specific understanding and interpretation of weathering rates and losses.

6. References

1. Dragovich D., Weathering rates of marble in urban environments, eastern Australia, *Zeitschrift für Geomorphologie*, **N.F.30** (1986), 203-214.
2. Attewell P.B. and Taylor D., Time dependent atmospheric degradation of building stone in a polluted environment, *Engineering Geology of Ancient Works, Monuments and Historical Sites*, ed. Marinos G. and Koukis G. (1988), 739-753.
3. Cooke R.U. Inkpen R.J. and Wiggs G.F.S., Using gravestones to assess changing rates of weathering in the United Kingdom, *Earth Surface Processes and Landforms,* **20** (1995), 531-546.
4. Cooke R.U., Geomorphological contributions to acid rain research: studies of stone weathering, *Geographical Journal*, **159** (1989), 361-366.
5. Shelford A. Inkpen R.J. and Payne D., Spatial variability of weathering on Portland stone slabs, In *Processes of urban Stone Decay,* ed. Smith B.J. and Warke P.A. (1996), 98-110.
6. Gabriel G. and Inkpen R.J., The nature of decay of Monk's Park limestone under simulated salt weathering, *Proceedings of the 8th International Congress on Deterioration and Conservation of Stone*, ed. Riederer J. (1996), 573-578.
7. Shelford A. and Inkpen R.J., The assessment of geological structures as constraints in stone decay, In *Proceedings of the 8th International Congress on Deterioration and Conservation of Stone*, ed. Riederer J. (1996), 253-264.
8. Klein M., Weathering rates of limestone tombstones measured in Haifa, Israel, *Zeitschrift für Geomorphologie*, **N.F.28** (1984), 105-111.
9. Silk J., Analysis of covariance and comparison of regression lines, *CATMOG 20.*

7. Acknowledgements

The author would like to thank Jon Jackson and Pat Jackson for aid in collecting the data for Oxford, Clacton-on-Sea and their rural counterparts; and Pete and Gill Mallett for their hospitality during the collection of the Birmingham data.

Stone Weathering and Atmospheric Pollution Network '97: Aspects of Stone Weathering, Decay and Conservation.
Edited by M.S. Jones & R.D. Wakefield © 1998 Imperial College Press.

DECIPHERING THE IMPACTS OF TRAFFIC ON STONE DECAY IN OXFORD: SOME PRELIMINARY OBSERVATIONS FROM OLD LIMESTONE WALLS

S.J. ANTILL, H.A. VILES
Urban Environments Research Group, School of Geography,
University of Oxford, OX1 3TB, England.
Tel. 01865 271 919. Fax 01865 271 929
E mail heather.viles@geography.oxford.ac.uk

Many roadside walls within the historic city centre of Oxford have a long history of blackening and decay, and in several cases there has been an increase in visible damage over recent years. This may be related directly or indirectly to rising traffic levels. Detailed study of part of the limestone walls fronting Worcester College, near a busy road, reveals a diversity of weathering features dominated by a suite of blackened gypsum-rich crusts with evidence of a developmental sequence. Chemical analyses of a range of samples from different types of crusts at different heights on the wall indicates their relative chemical homogeneity. Lead levels increase towards the base of the wall, implicating traffic exhausts as a source, but high lead levels are not found at the level of maximum decay (where blisters and blowouts are concentrated). Further work is needed to elucidate the role traffic exhausts play in influencing deterioration on such encrusted walls.

1. Introduction

Many parts of the architectural heritage of Europe and elsewhere continue to be affected by air pollution and other damaging agents, despite general reductions in smoke and sulphur dioxide concentrations in urban atmospheres over the past 50 or so years. Several hypotheses have been put forward to explain such continuing decay. Firstly, motor vehicle traffic fumes have been proposed as a major cause. Exhaust emissions contain a range of corrosive gases (including nitrogen oxides and sulphur dioxide) as well as particulate matter rich in sulphur and heavy metals, all of which may contribute to stone decay. Rodriguez-Navarro and Sebastian [1], for example, have shown in laboratory experiments that carbon-rich and metal-rich particulates from vehicle exhausts are capable of enhancing limestone sulphation. Secondly, traffic may contribute indirectly to stone decay along roadside walls in locations where road de-icing salts are applied in winter months. Salts are known to be powerful agents of limestone weathering [2]. Thirdly, changing urban air quality may encourage microflora (e.g. lichens, algae, fungi and bacteria), whose growth may be enhanced by increased nitrogen dioxide levels, which themselves can have a powerful biodeteriorating effect on the underlying stone [3]. Finally, several authors have suggested that there may be an important 'memory effect' whereby pollutants deposited within porous stone in the past may become 'reactivated' under present conditions.

Untangling these various hypotheses has proved to be very difficult in real situations (in comparison with controlled laboratory experiments on fresh stone), but it is vital for present and future management of historic stonework that we understand how today's environment affects old stone. Thus, many recent studies have attempted to use geochemical evidence from such old walls to interpret the mechanisms of decay. For example, Vleugels *et al.* [4] sampled weathering crusts, rainfall and runoff waters on the Jeronimos Monastery, Lisbon and performed a range of chemical analyses in order to make inferences about the progress of decay on the Lioz limestone in this 16th century monument. In a similar study, Leysen *et al.* [5] studied crusts and runoff waters on St Rombouts Cathedral, Mechelen, Belgium which was built between the 13th and 15th centuries of sandy limestone. More recently, Duffy and Perry [6] sampled Portland Limestone from different parts of the Public Theatre, Trinity College Dublin, and used petrological and chemical observations to infer mechanisms of decay. These studies, and many other similar ones, have often found difficulties in sampling such fragile encrusted stone and it is sometimes unclear exactly what has been sampled from where.

Studies searching for evidence from decayed stone need to consider carefully the spatial pattern of weathering, as well as the history of weathering, in order to produce clear and meaningful data [7]. Most historic monuments are architecturally complex, incorporating a range of surface geometry, stone type and orientations, which lead to a clear spatial zonation of weathering forms. In Vleugel *et al.*'s [4] study of the Jeronimos Monastery, for example, the authors noted on the most deteriorated parts of the building the presence of thin brown crusts on areas with no exposure to rainfall, black crusts on infrequently washed surfaces and white eroded surfaces on frequently washed surfaces. They sampled from each of these areas, but make no detailed observations of the extent and heterogeneity of these forms. Several studies have been made which attempt to quantify and map the distribution of weathering forms on such historic monuments, as exemplified by the work of Fitzner and Kownatzki [8] at Petra, Jordan and elsewhere. However studies of this type have only rarely been carried out alongside detailed geochemical analyses. Where ancient stones are suffering from decay, their history, in terms of past treatments, exposure to varying types and levels of pollutants, can be as important as the current situation in explaining (and predicting) the nature and rate of decay in circumstances where a 'memory effect' is likely to be operating (e.g. on infrequently washed areas where blackened, gypsum-rich crusts build up). Compiling a history of the stones can be very difficult, but for many important historical monuments good records of management techniques may be found, and it is often possible to piece together a record of the progress of decay from old photographs.

In order to untangle the influence of traffic on the decay of historic buildings within an urban centre such as Oxford a combination of field survey, laboratory analyses and historical investigations on, ideally, similar walls both near roads and well away from traffic should be carried out. This paper reports on the early stages of such a project, where the spatial variability of decay forms and chemistry of the weathered crust along one wall has been observed and its implications for the role of traffic in stone decay made.

2. Study site

Oxford is an important academic, commercial and tourist centre in central England, with a large stock of historical monuments built largely out of limestone. Fifty years ago (in 1947) the great geologist W. J. Arkell produced a book entitled 'Oxford Stone' which provides an excellent introduction to the building stones and their decay problems. Arkell stated that in Oxford in 1947 it was "...necessary to seek out examples in which the decay of stone can be studied" because restoration and re-facing in the preceding 80 years had rendered decaying stonework rare. However, a large programme of restoration was required again in the 1960s and 1970s [9]. Despite drastic decreases in atmospheric smoke and sulphur dioxide over the last 30 years (mean annual smoke levels down from 44μg/m^3 in 1965 to 6μg/m^3 in 1995, and mean annual sulphur dioxide down from 30ppb in 1965 to 6ppb in 1995), Oxford stone is still decaying. Many people consider that the great increase in motor vehicle traffic within the city centre over the last few decades is at least partially responsible.

Worcester College, situated within the centre of Oxford on a busy road, provides a good example of an assemblage of historic buildings which have been partly restored, but where there is still much evidence of decay. Like many buildings in Oxford in the 17th century, the large part of Worcester College was built from local Headington limestone which (at least the freestone variety) has proved to be highly susceptible to decay in polluted air. Worcester College also contains some older buildings, dating from the 15th century, which are probably constructed from Wheatley limestone (another local limestone commonly used as a building material before Headington stone dominated). The front of Worcester College faces an extremely busy road junction, where Worcester Street, Walton and Beaumont Street meet. The wall investigated in this study faces east and shows patchily serious decay. It is approximately 20m long and at its lowest point, 2m high.

The broad history of the Worcester College wall is as follows. The ashlar walls on the east side of the chapel and dining hall were originally built in the early 18th century from Headington Stone and were partly refaced around the beginning of the 20th century, with further extensive re-facing in the 1960s in cream Clipsham Stone (a very durable limestone from Rutland). The rubble walls to the north and south of the ashlar wings originate from the 15th century. On the north side the wall first provides the east gable wall of the Senior Common Room, and this was most recently cleaned in 1983. Immediately north of this is a 15th century blocked-in gateway, in which has been reset a 16th century doorway, now also blocked [10]. The rest of the northerly section of this wall encloses the Fellow's Garden and it is this section which has been studied in detail here. Apart from the ashlar faces, the wall has only been patchily and sporadically repaired with new stone over the 20th century, and parts washed from time to time.

3. Methods for mapping decay features

A photographic survey was undertaken from which a base map was produced showing the position of each clearly visible stone block. Much of the upper middle section of the wall was so blackened and decayed, as well as partially obscured by overhanging vegetation, that individual stone blocks could not be identified. Each identifiable stone was numbered and all stones up to a height of around 2m were sampled. Constraints of time and access limited higher and more detailed sampling. For each stone block, two main types of data were collected. Firstly, the presence or absence of a list of decay features was noted (i.e. different types of crust - thick black, thin black, thick grey, thin grey and brown, dusty deposits, blisters and blowouts, organic growths). Secondly, using a simple transparent overlay divided into grid squares an estimate of the percentage cover of blackened and clean surfaces and blown-out zones was made. From this information maps of the distribution of decay features were produced, and simple statistical tabulations made. Overall, more than 700 blocks of stone were observed in this way. The intervening mortar was not sampled, although it too possessed a wide range of decay features and in many cases was more blackened than the surrounding stone blocks.

4. Methods for chemical analyses

4.1 Sampling strategy

Samples were collected on 9th and 10th May, 1997. In an effort to avoid marking the wall, sample volumes were kept small, especially where crust collection would involve removal of undamaged stone. Fifteen samples of each crust category were collected, five at approximately 1.8m above the pavement, five at approximately 1.0m and five at approximately 0.2m. Each group of five was spread as evenly as possible along the wall. The small number of samples taken at this preliminary stage limits the usefulness of any statistical calculations carried out on the results. Thus, it was considered more important to gain an overview of the wall chemistry along its entire length, rather than carry out stratified random sampling. A steel penknife was used to scrape the crust into plastic sampling bags, taking care to remove only the crust and not the underlying stone. An additional seven random samples were taken of thick and thin black and grey crusts for scanning electron microscopy (SEM) and energy dispersive analysis of X-rays (EDX) observations.

4.2 Sample preparation

Analyses by ion chromatography (IC) and atomic absorption spectrometry (AAS) require the removal of analytes from the solid crust into solution. The literature contains a wide range of alternative extractants and techniques, including sequential extractions, the use of chelators, acids and organic compounds, all of which remove different proportions of the analytes from the crust. The choice of extraction technique is important to the meaning of the results. Cost, speed and simplicity are also important factors. For the present study, single extraction with Milli-Q water was selected for the analysis of chloride, nitrate, sulphate and calcium [5, 6]. The amounts of sample and water were calculated so as not to exceed the cold water solubility limit of calcium carbonate (0.014g/L) or of gypsum (2.14g/L) [11].

For the analysis of lead and zinc, single extraction with 2M nitric acid was chosen. This technique is often used in the study of soils, and 2M nitric acid has been found to be one of the most effective extractors of heavy metals [12]. A problem with the use of acid is that it dissolves a large amount of calcium carbonate and calcium sulphate, creating a matrix high in calcium ions. These interfere with the AAS measurements, resulting in higher-than-expected values [13]. As a counter measure, a

2M nitric acid solution saturated with calcium carbonate was used as a blank. Since undoubtedly some of the extracts were undersaturated with respect to calcium, the results of the lead and zinc analyses are likely to be slight under estimations. It should be noted that calibration with calcium carbonate-saturated standards was unfeasible because of the unnecessary amounts of calcium that would thereby be introduced into the spectrometer. Calcium clogs the spectrometer tube and in sufficient quantities causes unreliability in subsequent measurements. Repeated aspiration of calcium-free acid reduces this problem, but since the enhancement of metal ion measurements by calcium ions is unrelated to the concentration of metal ions [13], subtraction of the interference due to a calcium carbonate-saturated blank was considered to be sufficient.

For the seven samples taken for SEM and EDX observations, two sample preparation techniques were employed. One part of each sample was fractured and mounted with epoxy resin on an aluminium SEM stub. The remainder of each sample was prepared by crushing, placing on a filter paper in a funnel and treating with 10% hydrochloric acid to remove most of the calcareous rock. Pieces of the filter paper were then mounted on aluminium stubs. All samples were observed with a Cambridge Stereoscan 90 with Link Systems EDAX.

4.3 Analyses

Six analytes were measured in stone extracts: chloride, nitrate, sulphate, calcium, lead and zinc. Chloride was used as it is an indicator for the presence of road de-icing salts, and nitrate and sulphate are indicators of deposition of atmospheric pollutants. Calcium was expected to be the most common analyte, and was included as a baseline with which to compare the concentrations of the others. Finally, lead and zinc were used as indicators of vehicle pollution [13, 14]. Zinc can be an indicator of waste incineration, however there are no known incineration activities carried out near the wall.

All glassware used was washed in 2M nitric acid then rinsed twice with distilled water and once with Milli-Q water. For the analyses of sulphates, nitrates and chlorides approximately 100mg of dried sample was sonicated (3hrs) in Milli-Q water (200mL). The process was repeated on random samples to ensure that three hours led to complete extraction. The extracts were filtered through Whatman 41 filter paper. After further filtration through a 0.2μm filter, the analyses were performed on a Dionex DX500 ion chromatograph. The extracts, after addition of 0.1% potassium chloride, were analysed for calcium on a Perkin-Elmer 3030 atomic absorption spectrophotometer according to the standard recommended method [15]. For analyses of lead and zinc contents, approximately 2g of dried sample was sonicated (1hr) in 2M nitric acid (Aristar, 30 mL). As above, the process was repeated on random samples to ensure that one hour of sonication led to complete extraction. The extracts were filtered through Whatman 41 filter paper. The analyses were performed on a Perkin-Elmer 3030 atomic absorption spectrophotometer according to the recommended method [15].

5. Results

5.1 Mapping decay features

Following the mapping exercise, three main surface decay features were observed as shown in Table 1. Table 2 indicates that the most commonly occurring forms of decay were hard thin grey crusts and unconsolidated dusty deposits. Figure 1 shows the combined distribution of all the different types of consolidated crusts, and Figure 2 shows the distribution of catastrophic decay (which includes both blistering of blackened crusts, and blow-outs which involve large scale removal of the surface crusts and much of the friable sub-surface stone). A clear spatial zonation of decay features was found, with blow-outs concentrated within a clear zone about 1-2m above pavement level.

Blackened crusts were more widely distributed, although dusty deposits were found more commonly near ground level. Similar zonation of decay is apparent on many roadside walls in Oxford, although the height of the 'blow-out zone' varies hugely, perhaps as a result of differences in microclimate and especially capillary rise. Table 3 shows a bimodal frequency distribution of blackening, with stones commonly showing blackening of 26-50% or 76-100% of their surface. There was widespread evidence of organic growths (mosses, lichens and fungi), with 10% of sampled stone blocks showing obvious growths. Organic growths were commonly observed on horizontal ledges within stone blocks, along the upper edges of stone blocks and within areas of mortar.

Table 1. Descriptions of decay features used in sampling.

Category	Description
Thin black crust	Thin, hard, black coating, appearing grey when less concentrated, for example when deposited on freshly exposed or washed stone
Blistering thick black crust	Thick, black crust forming a 'bubble' over heavily weathered stone
Dusty deposits	Light, grey deposit, easily removed without affecting underlying stone and usually found on unweathered and sheltered stone

Table 2. Occurrence of different crust types and decay features

Feature	% of blocks on which feature occurs (blocks may contain more than one feature)
Brown crust	5
Organisms (fungi, algae and moss)	10
Blisters and blowouts	13
Grey thick ropey crust	15
Hard thin black crust	22
Black thick ropey crust	23
Unconsolidated dusty deposits	30
Hard thin grey crust	38

Table 3. Distribution of percentage cover of blackening on individual blocks.

% cover of blackening	No. of blocks affected (% in brackets)
0-25	115 (16)
26-50	205 (29)
51-75	115 (16)
76-100	272 (39)

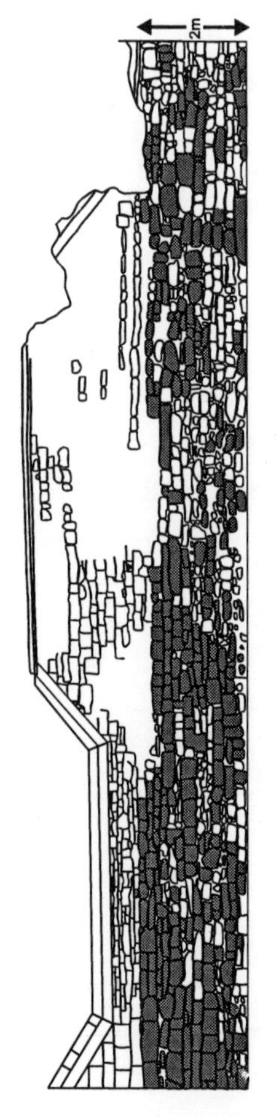

Figure 1 Distribution of hard thin grey and black, and thick ropey grey and black crusts.
N.B. The upper area of the wall (above 2 m) was not sampled in detail but showed a high degree of crust development.

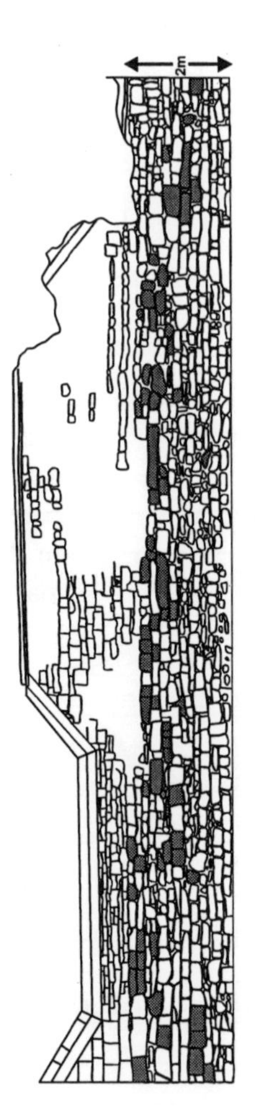

Figure 2 Distribution of blistered and blown-out stones.
N.B. The upper area of the wall (above 2 m) showed no significant blow-out development.

Evidence of past patching of the wall was also found, producing a range of stone types characterised by different types and degrees of decay. Much of the intensely blackened and blistered stone was found to be Headington Freestone (or a similar local stone) which was widely used during the building of the College and is known to react very badly to polluted air [16]. Blocks characterised by a hard, thin, grey crust were Headington Hardstone - which was again used widely within the College for plinths and other elements. Several blocks appeared to be Bath Stone, and others were blue-veined Clipsham which has been widely used for repairs and re-facings in the College during the 20th century. Clipsham blocks are notably durable, although often with grey discoloration.

Overlying this, in many places, were black crusts. Black crusts seemed to develop into thin layers on vertical faces, becoming thick and ropy in appearance within more sheltered areas (perhaps where redeposition can occur). Finally, there seemed to be a clear progression from grey to black crusts. Both have thin and thick ropey types, but in general the grey varieties are less hard and more friable. It would seem reasonable to suggest that the grey crusts represent an early stage in the development of black crusts. This was confirmed by observations on blown-out parts of stone where grey encrustations were noted to be developing on the new surface, but black crusts were only found on fragments of the old surface. A simple model of crust development drawn from these observations is presented in Figure 3.

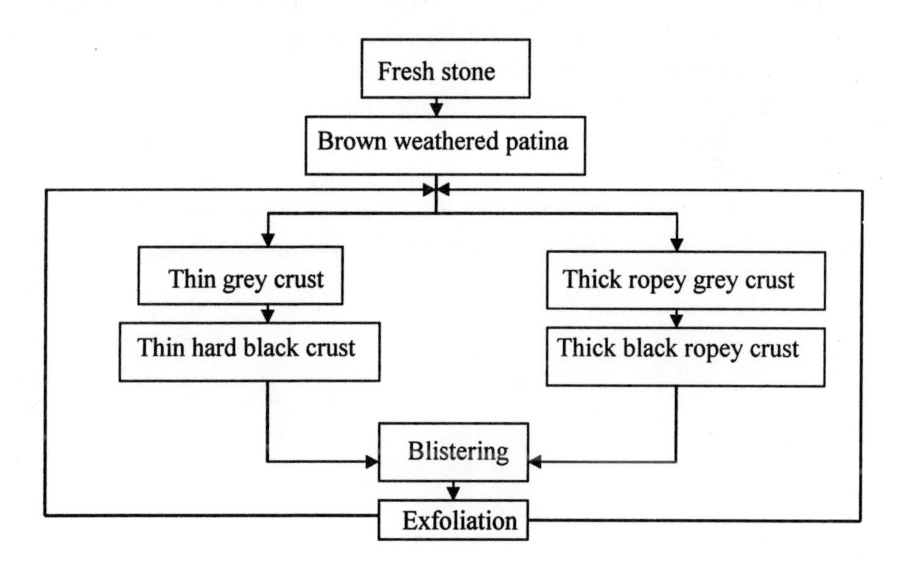

Figure 3. Model of crust development.

Across the wall, there was evidence for several generations of crust and there seemed to be a developmental sequence present. Thus, for example, clear patches of thin brown crust form a smooth, glassy surface, perhaps a natural weathering patina.

5.2 Chemistry of crusts

Repeat extractions of random samples showed sonication times to be sufficient. Table 4 gives the mean, standard deviation and range of concentrations for each analyte, Table 5 gives the correlation between analytes, Figure 4 shows the distribution of results according to height of samples, and Table 6 shows the correlation of concentrations and height above the pavement. These results compare well with other published data on the composition of crusts, with Leysen *et al.* [5] finding the same order of magnitude of chloride and nitrate in crusts from St Rombouts Cathedral, Mechelen, Belgium, although rather higher sulphate concentrations. The values for the heavy metals are in the same order of magnitude as those found by Bacci *et al.* [14] (Pb: >149ppm, Zn: 254ppm) and Camuffo *et al.* [17] (Pb: <1ppm-65ppm, Zn: 95-380 ppm). Student's t tests showed that sulphate was the only analyte that varied significantly ($p<0.05$) between the weathering features given in Table 1, with a much higher concentration in the blister samples.

The SEM and EDX observations revealed more about the micro-scale structure of the crusts and underlying rock. The untreated samples showed a large amount of gypsum within the white, friable zone below the harder blackened crusts. The samples treated with hydrochloric acid revealed a huge number of alumino-silicate particles (often spherical and porous, but with many smaller, spherical forms of around 20μm in diameter) characteristic of oil-fired and coal-fired combustion. Further analyses of the nature and distribution of these particles would help throw light on the role of traffic emissions in contributing material to crusts and facilitating stone decay.

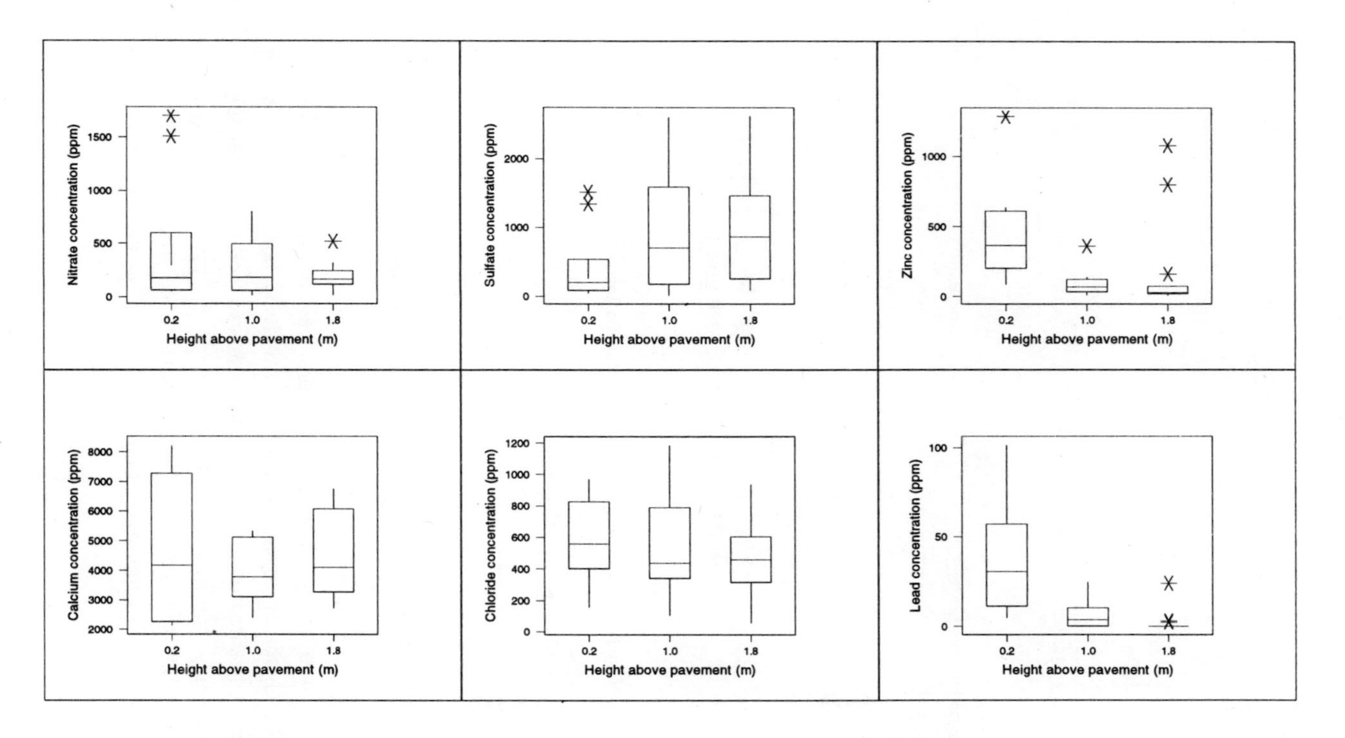

Figure 4 Concentrations of analytes in relation to height of samples above pavement level. The line drawn across the box represents the median value. The bottom of the box represents the first quartile (Q_1), the top of the box represents the third (Q_3) and the whiskers are drawn to the lowest and highest values in the range Q_1-1.5(Q_3-Q_1) to Q_1+1.5(Q_3-Q_1).

Table 4. Descriptive statistics of the analysis results (all values in ppm).

Analyte	Mean	Standard Deviation	Minimum	Maximum
Chloride	518	270	57	1186
Nitrate	286	374	16	1706
Sulphate	801	804	12	2627
Calcium	4404	1632	2112	8227
Lead	13	22	0	101
Zinc	224	316	7	1287

Table 5. Correlation between analytes.

	Chloride	Nitrate	Sulfate	Calcium	Lead
Nitrate	0.22				
Sulfate	-0.25	-0.35			
Calcium	0.11	0.24	0.2		
Lead	-0.10	-0.09	-0.35	-0.23	
Zinc	0.13	-0.09	-0.39	0.06	0.7

Table 6. Correlation of analyte concentration with height above the pavement.

	Chloride	Nitrate	Sulphate	Calcium	Lead	Zinc
r^2	-0.36	-0.27	0.27	-0.01	-0.63	-0.36

6. Discussion and conclusions

The results presented indicate that the great heterogeneity in appearance (colour and morphology) of weathering crusts on the front wall of Worcester College is not reflected in the chemistry of crusts, which is relatively homogeneous (except for a higher sulphate concentration in blisters) and compares well with other studied crusts. However, this study confirms observations made by previous authors about the difficulty of sampling such crusts without contamination from the underlying friable stone, especially where sampling is constrained (e.g. on historic monuments). The presence of lead on the Worcester College wall, especially at low heights above the pavement and in correlation with zinc, suggests that particles deposited on the wall from car exhaust fumes are becoming incorporated into the weathering crusts. However, the location of the zone of maximum decay (as represented by extensive blistering and blowouts) well above the area showing maximum lead contents indicates that, on this historic wall at least, there is no simple relationship between car exhaust emissions and limestone decay. It is likely that the lead and zinc are present in the heavier particles emitted from passing automobiles, so settle out of the air column rapidly, while the sulphur dioxide that causes gypsum formation travels further and higher. The data collected here on chloride contents does not confirm any suggestion that de-icing salts contribute to decay, although sampling during May (i.e. several months after any winter applications of road salt) has perhaps underestimated the presence of soluble chloride salts.

7. References

1. Rodriguez-Navarro C. and Sebastian E., Role of particulate matter from vehicle exahust on porous building stones (limestone) sulfation. *Science of the Total Environment* **187** (1996) 79-91.
2. Goudie A.S. and Viles H.A. *Salt weathering hazards*. (John Wiley, Chichester, 1997).
3. Koestler R.J. Charola A.E. Wypyski M. and Lee J.J., Microbiologically induced deterioration of dolomitic and calcitic stone as viewed by scanning electron microscopy. In *Proceedings, 5th International Congress on Deterioration and Conservation of Stone*, ed. Felix G., Lausanne, (1985), 617-26.
4. Vleugels G. Roekens E. Van Put A. Araujo F. Fobe B. Van Grieken R. Mesquita e Carmo A. Azevedo Alves L. and Aires-Barros L., Analytical study of the weathering of the Jeronimos Monastery in Lisbon. *The Science of the Total Environment* **120** (1992), 225-243.
5. Leysen L. Roekens E. and Van Grieken R., Air-pollution-induced chemical decay of a sandy-limestone cathedral in Belgium. *The Science of the Total Environment* **78** (1989), 263-87.
6. Duffy A.P. and Perry S.H., The mechanisms and causes of Portland Limestone decay - a case study. In: Riederer J. (ed.) *Proceedings, 8th International Congress on deterioration and Conservation of Stone*, Berlin, (1996), 135-45.

7. Halsey D. P. Dews S. J. Mitchell D. J. and Harris F. C., The black soiling of sandstone buildings in the West Midlands England: regional variations and decay mechanisms. In: *Processes of Urban Stone Decay,* ed. Smith B. J. and Warke P. A., Chapter 5. (Donhead Publishing, London, 1996), 53-65.
8. Fitzner B. and Kownatzki R., Kartierung und empirische Klassifierung der Verwitterungsformen und Verwitterungsmerkmale von Natursteinen an geschädigten Bauwerkspartien. *Bautenschutx + Bausanierung Sonderheft 'Baususbstanzerhaltung in der Denkmalpflege',* Köln: Verlagsgesellschaft Rudolf Müller GmbH, (1989), 21-25.
9. Oakeshott W.F., ed. *Oxford Stone restored.* (Trustees Oxford Historic Buildings Fund, Oxford, 1975).
10. Royal Commission on Historical Monuments, *The City of Oxford* (H.M.S.O. London, 1939) 124.
11. Weast R.C., (ed.) *CRC handbook of chemistry and physics* (CRC Press, 59th edition, 1978).
12. Desaules A. Lischer P. Dahinden R and Bachmann H. J., Comparability of chemical analysis of heavy metals and fluorine in soils: results of an inter-laboratory study. *Commun. Soil Sci. Plant Anal.* **23** (3&4), (1992), 363-377.
13. Antill S.J., Interference in AAS measurements of lead in calcium-rich matrices. Case study: Jenolan Caves. BA Dissertation, Department of Inorganic Chemistry, University of Sydney, (1995), 52.
14. Bacci P. Del Monte M. Sabbioni C. and Zappia G., *Black crusts as air pollution indicators.* In *Science, technology and European cultural heritage.* ed. Baer, N. Sors, A. (Butterworth-Heinemann, Oxford, 1989), 462-4.
15. Perkin Elmer., *Analytical methods for atomic absorption spectrophotometry.* (1982).
16. Arkell W.J. *Oxford Stone.* (Faber and Faber, London, 1947) .
17. Camuffo D. Del Monte M. and Sabbioni C., Origin and growth mechanisms of the sulphated crusts on urban limestone. *Water, Air and Soil Pollution* **19**, (1983), 351-9.

8. Acknowledgements

We would like to thank the Sir Robert Menzies Centre for Australian Studies for financial assistance with this project, Mr Martin Stephen for field assistance and the Provost and Fellows of Worcester College for allowing us to study their wall.

Stone Weathering and Atmospheric Pollution Network '97: Aspects of Stone Weathering, Decay and Conservation.
Edited by M.S. Jones & R.D. Wakefield © 1998 Imperial College Press.

ROLE OF ATMOSPHERIC SULPHUR DIOXIDE IN THE SULPHATION REACTION OF FRESCOES

N. SCHIAVON
Department of Earth Sciences, University of Cambridge
Downing Street, Cambridge CB2 3EQ, England
Tel. ++44 (0) 1223-333400; Fax ++44 (0) 1223-333450
E mail ns126@esc.cam.ac.uk

G. SCHIAVON
Dipartimento di Chimica "G. Ciamician", Universita' di Bologna
via Selmi 2, 40126 Bologna, Italy
Tel. ++39 51 259471; Fax ++39 51 259456;
E mail schiavon@ciam.unibo.it

The weathering reactions and products occurring on the surface of a sample of painted plaster prepared using techniques and materials used by old masters in frescoes and exposed to a simulated polluted atmosphere under a continuous flow of SO_2+ H_2O have been investigated by FT-IR. Preliminary results indicate that the sulphation reaction on carbonate material in frescoes involves two distinct reactions (Lewis acid/base reaction and redox reaction), and that the formation of the final sulphate product is always preceded by an intermediate sulphite stage.

1. Introduction

As part of a wider investigation on the role of sulphur dioxide (SO_2) as a weathering agent on monument and works of art, we have here focused our attention on sulphation affecting mural paintings in an indoor environment. The sulphation reaction in frescoes leading to the precipitation of gypsum ($CaSO_4 \cdot 2H_2O$) on the painted plaster's surface is a well-known decay phenomenon and has been the subject of several research papers in conservation science in recent years [1-5]. Despite the wealth of both experimental and observational data gathered, there is still little evidence and consensus on the detailed chemical pathway of the reaction leading to the transformation of the calcium carbonate ($CaCO_3$) in the plaster into calcium sulphate ($CaSO_4$) and the exact nature of the weathering products [6]. In order to provide answers to some of these questions, we have carried out research on sulphation of indoor frescoes combining experimental tests on unweathered plaster samples exposed to simulated sulphur dioxide attack with chemical, mineralogical and microscopic analyses (by Fourier Transform Infrared Spectroscopy {FT-IR} and Scanning Electron Microscopy {SEM} with Energy-Dispersive X-ray analysis. {EDX}) of weathering products present on plaster and pigment materials in ancient frescoes subjected to urban polluted atmosphere.

Experimental results from a sulphur dioxide attack test on a sample of painted plaster are presented in this work, and a sulphation reaction mechanism is proposed. The sample was prepared using the same techniques as those traditionally used by Italian masters centuries ago and it is composed of a grey "arriccio" in contact with the mural substrate and an overlying white "tonachino" painted on the surface with inorganic pigments.

2. Materials and Experimental

The "arriccio" is originally composed of lime (CaO which is readily transformed into $CaCO_3$ by atmospheric carbon dioxide) and coarse-grained silica sand (SiO_2) with a 1:2 ratio; the overlying "tonachino" is composed of lime, fine-grained silica sand and marble powder on a 1:1:1 ratio. The surface of the "tonachino" was coated with a green pigment derived from powdered malachite ($CuCO_3 \cdot Cu[OH]_2$). For the purpose of this study, the "arriccio" layer was carefully removed and only the "tonachino" layer with the green surface coating was used in the analytical routine. The sample of the "tonachino"(named sample 1a) was placed in a sintered glass funnel of fine porosity connected with a double-necked flask containing distilled water. Sulphur dioxide (SO_2) from a liquid gas cylinder (pressure = $3x10^5$ Pa) was funnelled through the second flask and was bubbled through water at a rate of 100 ml min^{-1}. A continuous flow of an aeriform mixture of SO_2 (g) and H_2O (vap) (hereafter referred to as 'gas') was then made to react with the "tonachino" sample. Artificial light from two visible lamps (100W) and an ultraviolet lamp (40W and λ=380nm) simulated solar illumination. The experimental conditions were designed to reproduce in a few hours weathering rates normally reached after several years of exposure to air pollution in urban areas. This set of conditions allowed the identification of experimental reactants and products and a reaction mechanism to be proposed.

Preliminary results from these experiments were obtained analysing the sample at fixed time intervals by using a Nicolet 205 FT-IR spectrometer. Infrared absorption spectra were obtained using KBr disks prepared by adding 1.5mg of pre-ground sample to 300mg of KBr (Aldrich FT-IR grade) and by pressing (9 tons for 4 minutes) to form a disk 13mm in diameter and 1.5mm thick. Powder from the sample was obtained by manually scratching with a blade the surface of the "tonachino" and although some contamination of surface samples from the underlying "arriccio" substrate composed of SiO_2 and $CaCO_3$ was inevitable care was taken to minimise it.

3. Results and Discussion

In the first experiment (Experiment A) only the white internal non-coated surface of the "tonachino" was analysed (i.e. opposite that of the green pigmented surface). Table 1 reports the wave number values for the characteristic absorption bands of four pure compounds, used here as reference standards, and corresponding to the "tonachino" composition ($CaCO_3$ SiO_2) and to the presumed weathering products after SO_2 attack ($CaSO_3 \cdot 2H_2O_2$ $CaSO_4 \cdot 2H_2O$). Table 1 also reports wave number values for the infrared absorption bands of three samples of white "tonachino" corresponding to the following reaction times: t_1=zero (sample 1a); t_2=1hr (3600s; sample 1b); t_3=3hr (10800s; sample 1c), i.e. before and after 1 and 3 hours of continuous exposure to the gas flow.

From the data in Table 1, using the method of positive and negative spectral interpretation [7], the following considerations can be made:
a) the spectrum of sample 1a presents the absorption bands of $CaCO_3$ and of SiO_2, i.e. of the constituents of the "tonachino".
b) the spectrum of sample 1b presents the absorption bands of SiO_2 of $CaSO_3 \cdot 2H_2O$ and of traces of $CaCO_3$ which did not react with the gas after an hour of flow.
c) the spectra of sample 1c presents only the absorption band of SiO_2 (not affected by the gas flow) and of $CaSO_4 \cdot 2H_2O$. After three hours of gas flow, surficial sulphation is completed and all the $CaCO_3$ has reacted.

Figures 1-3 present the infrared spectra of sample 1a, 1b and 1c. These spectra clearly show the progress of the sulphation reaction, which proceeds via the formation of $CaSO_3$ as the intermediate reaction product. In Table 1, wave number values corresponding to the absorption bands of the crystallisation water in $CaSO_3 \cdot 2H_2O$ and $CaSO_4 \cdot 2H_2O$ are not included; differences in bands shape, bands multiplicity and wave number values between the two dihydrated salts can be appreciated in Figures 2 and 3. In the spectrum of Figure 2, $CaSO_3 \cdot 2H_2O$ presents a unique band relating to the stretching vibrational mode of the O-H bond at 3398cm^{-1} and a unique band relative to the bending vibrational mode of the bond angle at 1629cm^{-1}; in the spectrum of Figure 3, $CaSO_4 \cdot 2H_2O$ presents two bands at 3549 and at 3407cm^{-1} relating to the stretching vibrational mode of the O-H bond and two bands at 1684 and 1621cm^{-1} relating to the bending vibrational mode of the bond angle. All of these bands are typical and unique

to gypsum. In this first analytical experiment, where only the white internal surface of the "tonachino" sample was analysed, the reaction
$CaCO_3 \rightarrow CaSO_3 \rightarrow CaSO_4$ is then complete.

Table 1 and spectra in Figures 1-3 reveal the importance of the use of infrared spectrophotometric analysis as a tool for investigating the progress of a chemical reaction. Careful interpretation of FT-IR spectra indicates unequivocally the absence and/or presence of certain reactants and products enabling us to obtain "snap-shots" of a reaction in progress.

Table 1. FT-IR data (experiment A). Wavenumbers in cm^{-1}. vs= very strong; s = strong; m = medium; br = broad; sp = sharp.

$CaCO_3$	1421vs (br)					875vs (sp) 712s				
SiO_2		1166s		1086vs			798vs 779s 695m			
$CaSO_3 \cdot 2H_2O$					982vs 942vs				652vs	
$CaSO_4 \cdot 2H_2O$			1143vs 1117vs					670m		602vs
Sample 1a	1433vs	1164s		1089vs			798m 779m 695w			
Sample 1b	1430m	1165s		1086vs	987vs 943s	875s (sp) 713w	798s 779s 695w		654m	
Sample 1c			1144vs 1118vs	1090vs			798s 779s 694m	669m		602s

The experimental evidence presented suggests that the sulphation mechanism leading to the transformation of carbonates into gypsum operate in two distinct stages and following two different reaction types. The first stage is an acid/base Lewis reaction of the type:
adduct (1) + acid (2) → acid (1) + adduct (2) which in our case can be written

$$CO_3^{2-} + SO_2 \rightarrow CO_2 + SO_3^{2-} \quad (1)$$

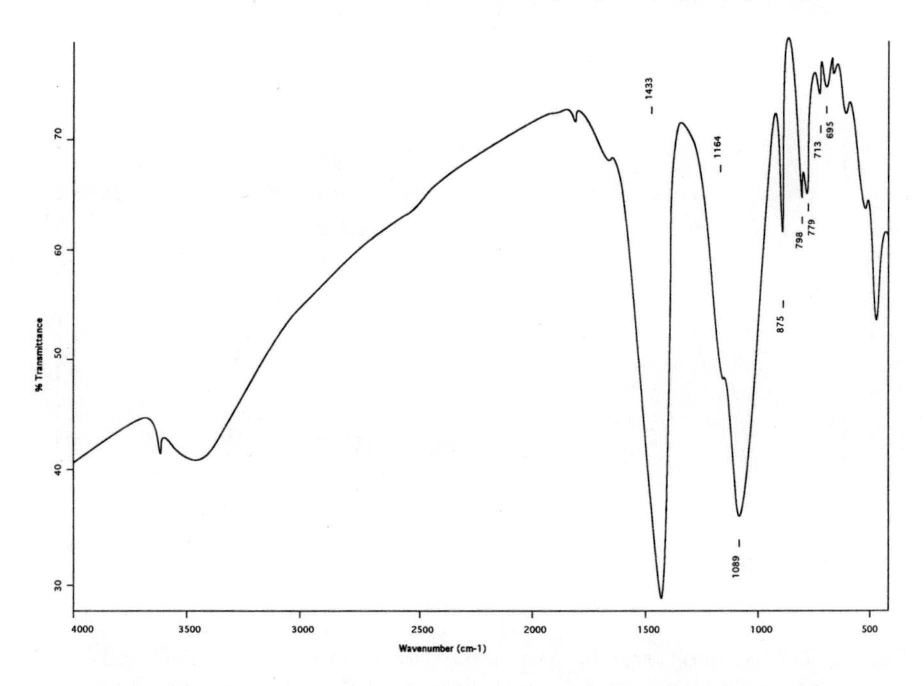

Figure 1. FT-IR spectrum of sample 1a. White surface of the "tonachino" before gas flow.

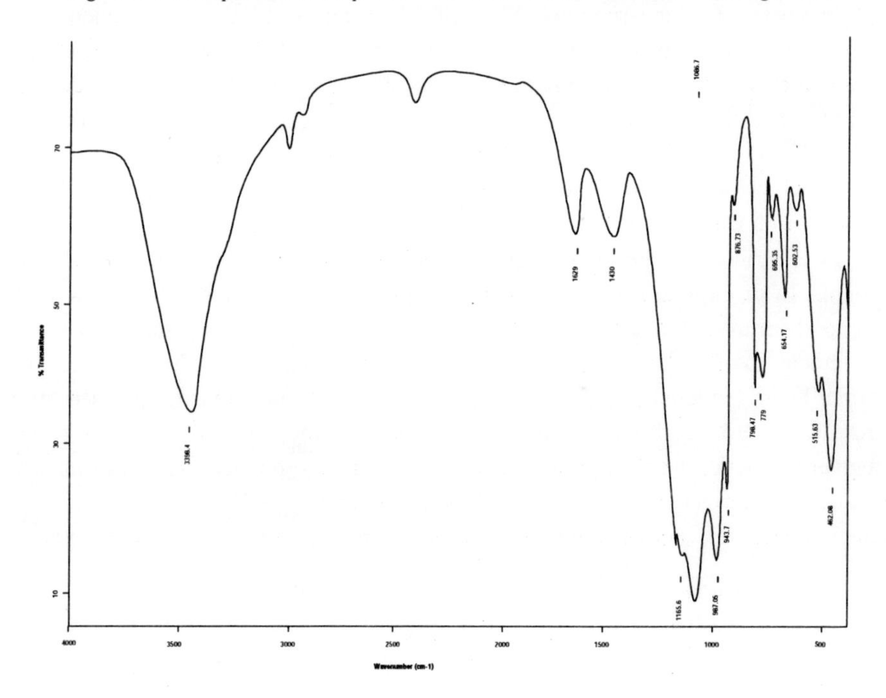

Figure 2. FT-IR spectrum of sample 1b. White surface of the "tonachino" after 1 hour of gas flow.

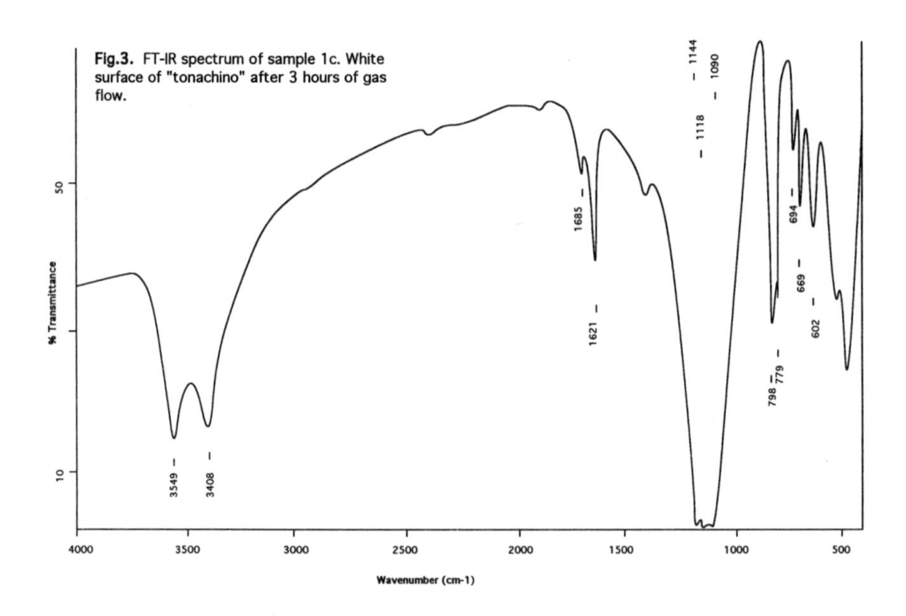

Figure 3. FT-IR spectrum of sample 1c. White surface of the "tonachino" after 3 hours of gas flow

In reaction (1) the strong acid SO_2 shifts the weak acid CO_2 from its adduct CO_3^{2-} forming in turn its own adduct SO_3^{2-}. It is worth remembering that, according to Lewis, the two anions (CO_3^{2-} and SO_3^{2-}) are acid/base adducts between the acids (CO_2 and SO_2) and the base (O^{2-}). The second stage of the reaction is the oxidation of the $CaSO_3{\cdot}2H_2O$ to $CaSO_4{\cdot}2H_2O$. In our case, in the absence of air, the oxidation is carried out by SO_2, which in turn is reduced to yellow elemental Sulphur, as confirmed by the presence of a yellow deposit on the surface of sample 1c.

In experiment B the green pigmented surface of the "tonachino" was analysed. Sample 1c of experiment A was used. Sample 1c was selected because no visible colour alteration had occurred on the green surface after exposure to the three hours of continuous gas flow of the previous experiment. The sample was then exposed to ten further hours (36000s) of gas flow after which a change in colour from green to red-brown occurred. At this point, the experiment was stopped. Table 2 presents wave number values for the absorption bands typical of three reference standards (malachite = $CuCO_3{\cdot}.Cu\ [OH]_2$), the reagent; $CuSO_3.{\cdot}2H_2O$ and $CuSO_4.5H_2O$, the expected reaction products) and wave number values for the absorption bands which appear in the spectra of three samples corresponding to: sample 2a = green surface of sample 1c before the extra ten hours of gas flow; sample 2b = red-brown surface after ten extra hours of gas flow; sample 2c = blue powdery residue obtained by water evaporation of the solution in the two necked flask at the end of experiment B.

From the data in Table 2 the following considerations can be made:

a) the spectrum of sample 2a presents all the absorption bands of $CuCO_3{\cdot}Cu[OH]_2$ + other bands which can be attributed to contamination from compounds (SiO_2 and $CaSO_3{\cdot}2H_2O$) of the underlying "tonachino" exposed to three hours of gas flow in experiment A.

b) the spectrum of sample 2b shows the presence of the absorption bands of $CuSO_3$ $2H_2O$ in the red-brown surface. Contamination in the form of SiO_2 and $CaSO_3 {\cdot}2H_2O$ is still present but the presence of band doublets in the range 3600-3400cm^{-1} and 1700-1600cm^{-1}, typical of crystallisation water of gypsum, points to the presence of $CaSO_4{\cdot}2H_2O$.

c) the spectrum of sample 2c shows the presence in solution of all the absorption bands typical of the water-soluble salt $CuSO_4{\cdot}5H_2O$.

A more detailed analysis of the spectra presented in Figures 4-6 leads us to the following observations. In the spectrum of Figure 4, the absorption bands corresponding to $CuCO_3{\cdot}Cu[OH]_2$, SiO_2 and $CaSO_3{\cdot}2H_2O$ are present. $CuCO_3{\cdot}Cu(OH)_2$ presents a doublet at 3406vs, 3315s cm^{-1} relating to the stretching mode of the O-H bond in the OH^- ions and absorption bands at 1492vs, 1442vs, 1394vs, 1089v, 1053s, 876vs, 821, 751, 712w cm^{-1}. The band at 1089cm^{-1} appears to be shifted in the spectrum (to 1097cm^{-1}) because of merging with the 1086cm^{-1} band due to SiO_2. SiO_2 presents absorption bands at 1166s, 1986vs, 798s, 779s, 695m cm^{-1}. $CaSO_3{\cdot}2H_2O$ presents absorption bands at 1628m, 982s, 946s, 654m cm^{-1}.

The presence of $CaSO_3$ underlying the green pigmented surface suggests that the $CuCO_3{\cdot}Cu[OH]_2$ layer coating the "tonachino" is not able to prevent SO_2 attack on the $CaCO_3$ substrate while it may slow down subsequent oxidation from $CaSO_3{\cdot}2H_2O$ and $CaSO_4{\cdot}2H_2O$ i.e. calcium sulphite to gypsum.

In the spectrum of Figure 5, obtained analysing the red-brown surface after ten hours of gas-flow, the absorption bands corresponding to copper sulphite, silica, calcium sulphite and calcium sulphate can be seen. $CuSO_3{\cdot}2H_2O$ presents absorption bands at 1249m, 1197s, 1136vs,1124vs, 1094vs, 994s cm^{-3}. SiO_2 presents absorption bands at 1094vs, 798s, 779s, 695m cm^{-3}. $CaSO_3{\cdot}2H_2O$ presents absorption bands at 947m^{-s}, 656m cm^{-1} while the band at 982 cm^{-1} is here masked by the 1094vs band of copper sulphite.

Table 2. FT-IR data (experiment B). Wave numbers in cm^{-1}.

Malachite $CuCO_3$ $Cu(OH)_2$	1492vs 1421vs 1390vs							1097s	1048vs		876vs 821vs		750s 712m					
$CuSO_3$ $2H_2O$		1252m		1197m			1135s 1124vs	1095vs		994m							618m	
$CuSO_4$ $5H_2O$			1202vs			1159vs		1098vs		996vs 961s					660vs			603s
Sample 2a	1491s 1422s 1394s				1166s			1089vs	1053s	991s 946m	876m 821m	798s 779s 695m	751m 712w			654m		
Sample 2b		1249m		1197s			1135s 1124vs	1094vs		994s 947m		798s 779s 695m		670m		656m		603s
Sample 2c						1160vs		1097vs		996vs 961s					661s			603s

In Figure 5, the intensity of the bands corresponding to calcium sulphite may qualitatively indicate that the abundance of $CaSO_3{\cdot}2H_2O$ in sample 2b is not high. $CaSO_4{\cdot}2H_2O$ presents absorption bands at 3544vs, 3406vs, 1184w, 1623m, 670w, 603m cm^{-1}; typical gypsum bands at 1144, 1117 cm^{-1} are not present due to interference from other compounds with bands in the range 1100-1200 cm^{-1}.
A possible explanation for the presence of gypsum in sample 2b is that the thinner and more porous red-brown surface layer (as opposed to the green pigmented layer of sample 2a) does not prevent the oxidation of sulphite to sulphate (i.e. $CaSO_3{\cdot}2H_2O$ to $CaSO_4{\cdot}2H_2O$), not withstanding also the ten extra hours of gas-flow.

In the spectrum of Figure 6, only absorption bands relating to $CuSO_4{\cdot}5H_2O$ are present. In detail, absorption bands at 3350vs, vbr and at 1625m cm^{-1} corresponding to the stretching and bending vibrational mode of the O-H bonds of the crystallisation water and at 1201vs, 1160vs, 1097vs, 996vs, 961s, 661s,603s cm^{-1} can be seen.

These results suggest that, as it is case with calcium carbonate, the sulphation of copper basic carbonate occurs in two distinct stages following two distinct reaction mechanisms: acid/base of Lewis reaction followed by redox reaction by SO_2 on a copper sulphite intermediate.

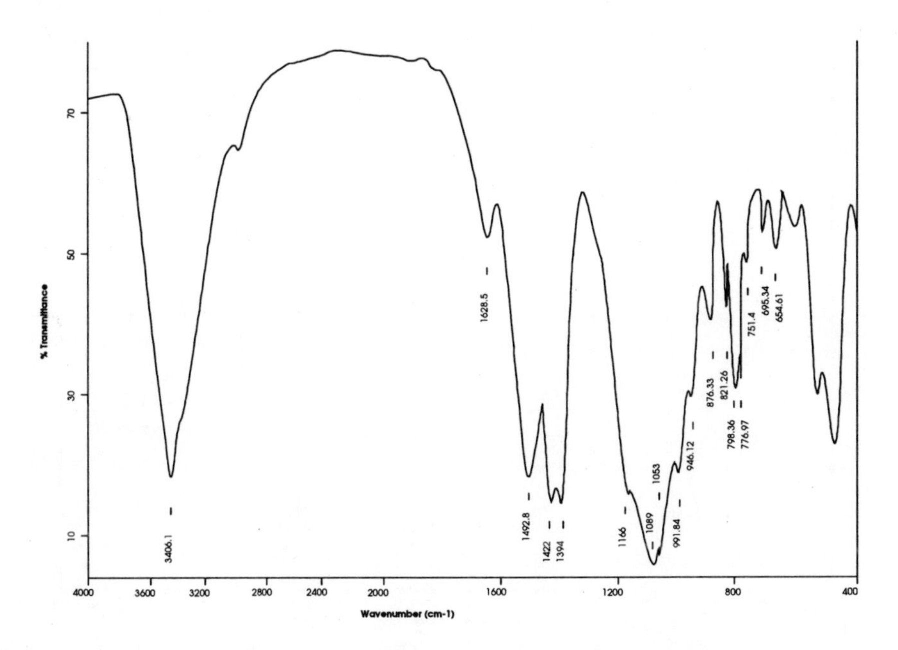

Figure 4. FT-IR spectrum of sample 2a. Green surface of the “tonachino” after 3 hours of gas flow.

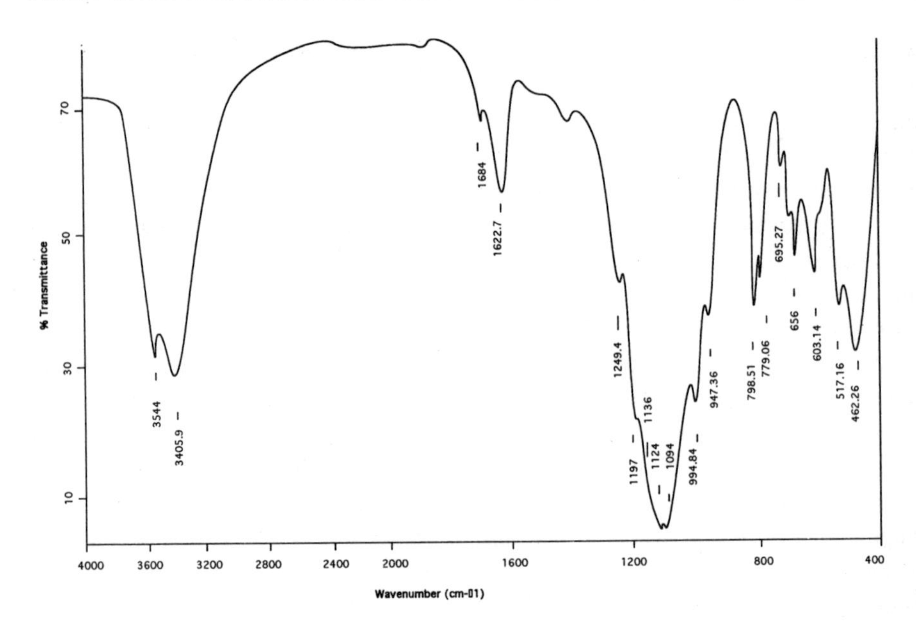

Figure 5. FT-IR spectrum of sample 2b. Red-brown surface after 10 hours of gas flow.

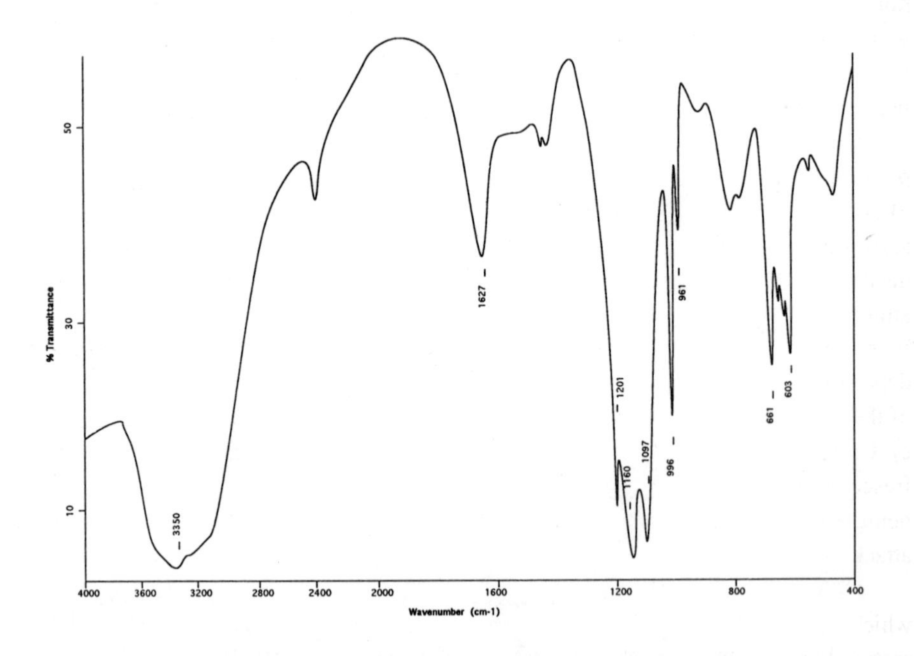

Figure 6. FT-IR spectrum of sample 2c. Residue obtained from aqueous solution (in two necked flask) by water evaporation after 10hours of gas flow.

4. Conclusions

Experimental attack by an aeriform mixture of SO_2 (g) and H_2O (vap) on a sample of plaster painted with green malachite $CuCO_3 \cdot Cu[OH]_2$ powder has shown that the final weathering sulphate product forms via oxidation of an intermediate sulphite stage.

In experiment A the following reaction path has been shown:

$$CaCO_3 \rightarrow CaSO_3 \rightarrow CaSO_4$$

the reaction above occurs on the white surface of the "tonachino" in two distinct phases: the first stage involves the attack by the Lewis acid on the acid/base adduct CO_3^{2-} and takes 1 hour to complete; the second stage involves the oxidation of SO_3^{2-} to SO_4^{2-} with the reduction of SO_2 to elemental S, and is completed in two hours.

In experiment B, the following reaction has occurred:

$$CuCO_3 \cdot Cu\,[OH]_2 \rightarrow CuSO_3 \cdot 2H_2O \rightarrow CuSO_4 \cdot 5H_2O$$

Also in this case, the sulphation proceeds via the formation of an intermediate sulphite compound but there are differences with experiment A. In A, the two stages of the reaction are distinct and occur at different, well defined times after 1 and 2 hours of exposure to the simulated polluted conditions. This is not the case in Experiment B. From Table 2, it can be seen that the spectrum of sample 2a after 3 hours of gas flow still presents all the absorption bands of $CuCO_3 \cdot Cu[OH]_2$ malachite which does not seem to have been altered by the gas attack. Only after an extra 10 hours of exposure time can the colour change to red-brown and the contemporaneous presence in the infrared spectrum of copper sulphite and copper sulphate be detected.

Ongoing and future research stemming from the results presented in this preliminary paper will continue along the following lines.

a) samples of "tonachino" at the intermediate "sulphite stage" will be exposed to an aeriform mixture composed by SO_2 (g) + H_2O (vap) and O_2 (g) to assess the time needed to achieve the above mentioned oxidation from sulphite to gypsum by atmospheric oxygen.

b) samples of pigmented plaster will be subjected to a direct SO_2 gas-flow (dry-deposition) to experimentally confirm the Lewis acid/base reaction as the first phase of the sulphation reaction of $CaCO_3$.

c) Comparative FT-IR spectrophotometric analysis of heavily weathered ancient frescoes samples both in indoor and outdoor (i.e. cloisters) environments is currently being carried out to assess the presence of "alien" components from atmospheric attack.

FT-IR analyses are being complemented by SEM + EDX chemical investigation which is proving particularly useful when studying ancient pigments composed of metal oxides which present very broad and difficult to interpret FT-IR bands.

5. References

1. Fassina V. Rossetti N. Zucchetta E., Mural painting in S. Tarasio chapel in the S. Zaccaria church, Venice: a study of the state of conservation of paint film. In *Proc. 8th Int. Congress on Deterioration and Conservation of Stone,* ed. Riederer J., Berlin (1996), 1663-1668.
2. Negrotti R. Realini R. Toniolo L., The frescoes of the Oratory of S. Stefano (Lenate sul Seveso - Milan): comparison between indoor air quality and decay. In *Proc. 8th Int. Congress on Deterioration and Conservation of Stone*, ed. Riederer J., Berlin (1996), 1647-1653.
3. Toniolo L. Cariati F. Polesello S. Pozzi A., The mural paintings in Melzo (Italy): analytical investigation of the pigmented layers. In *Proc. 8th Int. Congress on Deterioration and Conservation of Stone,* ed. Riederer J., Berlin (1996), 1655-1661.
4. Bacci M. Casini A. Lotti F. Piccollo M. Radicati B. Stefani I., La spettroscopia d'indagine nello studio preliminare di dipinti. *Abstract of Symposium:Conservazione e Restauro dei Beni Culturali. Turin.* (1996).
5. Marabelli M., Surveys for the control of indoor pollution affecting cultural property in Italy. In *Proc. of 3rd Int. Conference on non-destructive testing, microanalytical methods and environmental evaluation for the study and conservation of works of art.* (1992), 1026-1050.
6. Elfving P. Johansson L.G. Lindqvist L., A study of the sulphation of silane-treated sandstone and limestone in a sulphur dioxide atmosphere. *Studies in Conservation* **39** (1994), 199-209.
7. Socrates G., *Infrared Characteristic Group Frequencies,* 2nd. edition. (John Wiley & Sons, New York, 1994).

6. Acknowledgements

We thank Prof. Enzo Ferroni, a world leading figure in the field of frescoes restoration, from the Department of Chemistry at the University of Florence, for useful discussions and for providing the sample of painted plaster used in this study.

Stone Weathering and Atmospheric Pollution Network'97: Aspects of Stone Weathering, Decay and Conservation.
Edited by M.S. Jones & R.D. Wakefield © 1998 Imperial College Press.

MOISTURE LOSS FROM STONE INFLUENCED BY SALT ACCUMULATION

B.J. SMITH, E.M. KENNEDY
School of Geosciences, The Queen's University of Belfast,
Belfast, BT7 1NN, UK
Tel. 01232 335144. Fax 01232 321280
E mail b.smith@qub.ac.uk
E mail e.kennedy@qub.ac.uk

Experiments are described which quantify moisture loss over 24 hours from Portland Limestone and Baumberger Sandstone treated with 10% Na_2SO_4 and $MgSO_4$ solutions in a climatic cabinet at 38°C. Preliminary results show that after only three cycles of wetting and drying consistent differences occur between patterns of moisture lost related to the accumulation of salts. This highlights the dangers of explaining the results of salt weathering simulations and durability tests strictly in terms of an initial wetting and drying regime and the number of times that this has been repeated. At the end of each heating cycle all blocks experienced a short-lived period of rapid moisture loss as the temperature of the air fell more rapidly than that of the block. Such conditions occur naturally at for example, sunset, and if salt solutions within a stone are near to saturation this evaporation could lead to rapid crystallisation that might contribute to stone decay.

1. Introduction

Salt-rich environments include hot and cold deserts, coastal regions and, in particular, polluted urban and industrial areas. In polluted environments salt-induced decay is the key process by which many stone buildings and structures are damaged and disfigured. The mechanisms that operate and the decay features produced clearly depend upon a number of exogenic factors (principally temperature and moisture regimes and salt availability) and endogenic or 'stone' factors including thermal properties, physical strength and porosity/permeability characteristics. Porosity is especially significant due to the role of pore size and connectivity in determining moisture movement, and the significance of microporosity in controlling the ability of crystallised salts to bridge pore spaces and hence weathering susceptibility [1]. In terms of exogenic controls, the rate of drying of wetted stone containing salts is considered to exert a significant influence on salt accumulation and patterns of decay. Under conditions of rapid drying (high surface temperatures, surface winds and low relative humidity) surface layers may dry out more quickly than salt in solution can be drawn through the stone from below. The effects of this are thought to be the initial precipitation of some salt on, or immediately below, the surface and a further accumulation near to a wetting front which crystallises as water vapour is lost [2]. This accumulation of salt at depth is thought to be responsible for contour scaling observed on stonework [3]. Alternatively, slow drying of stone is more likely to allow salt in solution to be brought to the surface and to produce the granular disintegration or 'sanding' that is characteristic of humid, shaded areas on buildings [4].

Many of the above assumptions about salt-induced decay and its controlling factors are derived from field observations and conceptual models with little if any data on the actual wetting and drying regimes experienced by stonework. These models have in turn been tested by laboratory simulation experiments designed to replicate weathering mechanisms under controlled conditions, and/or to compare the durability of different stone types and the aggressiveness of salts and salt combinations. In such tests it is invariably assumed that wetting and drying regimes within test blocks remain constant over the duration of the experiment, despite the regular addition of more salt and/or the relocation of salt already within blocks. This assumption is questionable, as salt already within a block is re-mobilised during each wetting phase and modifications to pore structure and permeability produced by salt deposition are likely to influence the rate and depth of moisture movement into the stone. In extreme cases crystallised salt at and just below the stone surface has been seen to act as a 'passive pore filler', effectively preventing moisture ingress in arid environments [5]. This might also occur in sheltered areas of salt accumulation on buildings, where moisture is available in limited quantities from, for example, dew.

It might be expected therefore that as laboratory simulations, durability tests and natural weathering proceed, the wetting and drying characteristics of stone will vary, even if temperature regimes and amount and mode of presentation of salt and moisture remain constant. These changes should be conditioned by the design of the experiment, the initial porosity of the stone and the salt type used. This paper presents the results of preliminary experiments to examine the effects of porosity and salt type on stone moisture loss during the initial stages of a salt weathering simulation.

2. Methodology

Moisture loss was monitored by measuring weight loss from test blocks of two stone types of differing porosity characteristics under controlled heating within a climatic cabinet.

2.1 Materials and preparation

Two blocks each were used of Portland, an oolitic limestone from southern England, and Baumberger, a calcareous sandstone from western Germany, both of which have been used extensively as building stones. Porosity characteristics of the two stone types were determined using mercury porosimetry. Test blocks consisted of 10cm long cores of 7.5cm diameter that were varnished on the sides and set into insulating jackets of expanded polystyrene to leave one exposed end of each core. This procedure was designed to reproduce as far as possible conditions of exposure on a building where heating and cooling and the addition and loss of moisture are directed vertically through one face [6]. The blocks were suspended one at a time inside the cabinet by a fine steel wire passing through a small diameter hole in the roof and attached to an electronic balance. The weight of each block and its insulating jacket were recorded automatically by a computer linked directly to the balance.

2.2 Experimental procedure

Solutions (5cm^3) of 10% Na_2SO_4 or 10% $MgSO_4$ were applied evenly to the exposed surfaces of oven dried blocks of each stone type by syringe. These salts were chosen because of their widespread use in durability tests (Na_2SO_4) and simulation experiments ($MgSO_4$) [7]. After allowing the solutions to soak into the surface for 30 minutes each block was placed inside the cabinet and the air temperature raised from 20°C to 38°C over a 30 minute period. Thereafter, air temperature was kept at 38°C for 24 hours with a constant relative humidity of 17.5%. After 24 hours air temperature was decreased to room temperature (20°C) over the period of 1 hour. The temperature of 38°C was chosen to accord with conditions that can be expected on many buildings exposed to direct sunlight without the risk of temperature-induced fracturing of the stone [8]. The weights of the blocks were measured at 10 minute intervals, commencing 30 minutes after the start of the heating cycle so as to allow any residual motion in the suspended blocks to subside. Due to the susceptibility of the blocks to external vibrations it was decided to plot weight loss data as a running mean of three readings. The full procedure was repeated three times for each of the four blocks to assess the early stages of loading blocks with crystallised salt prior to the onset of any structural damage.

3. Results and discussion

Mercury porosimetry measurements (Figure 1) show that the two stone types have distinctive pore size characteristics. The Portland limestone has a bimodal distribution with a major peak between approximately 30μm and 8μm and a secondary peak associated with micropores between 0.3μm and 0.08μm. This produces a total porosity in the samples used of 18%. Baumberger sandstone, on the other hand, has a unimodal pore size distribution with a strong peak between 1μm and 0.8μm and a total porosity of approximately 21%. The predominance of this microporous structure possibly reflects the highly interlocked nature of the crystalline calcite cement of this stone type.

Weight loss characteristics plotted against the square root of time are shown in Figures 2 and 3. The Baumberger sample treated with 10% $MgSO_4$ (Figure 2a) exhibits a trend that is common to all four test blocks, in that after 24 hours of heating most moisture was retained after the second cycle and least after the third cycle. The plot for the first cycle only shows a significant moisture loss after 40 minutes, which is followed by a smooth, exponential decrease over the subsequent 24 hours. During the second cycle, however, moisture loss begins more quickly, perhaps because of moisture that failed to infiltrate through salt crystallised near the surface during the first cycle. Following this rapid loss, the rate of decrease is less than that for the first cycle and consequently more moisture is retained after 24 hours. This lower rate of loss might reflect the closing of pore spaces near the stone surface and the 'trapping' of moisture that had penetrated more deeply into the stone. These patterns are accentuated in the third cycle, which shows a very rapid loss of surface/near surface moisture and, even though the subsequent rate of loss is less than that for the first cycle, the total moisture content remains less than for both previous cycles. Similar early phases of rapid moisture loss were observed for the second and third cycles of Baumberger with Na_2SO_4 (Figure 2b), but only some 40 minutes after measurements began. This might suggest that if moisture penetration were inhibited by existing salt within the block this took place at some depth below the surface. The absence of a near surface accumulation of salt might also explain the similar paths of all three drying curves and the close correlation between the final moisture loss figures for all of the cycles.

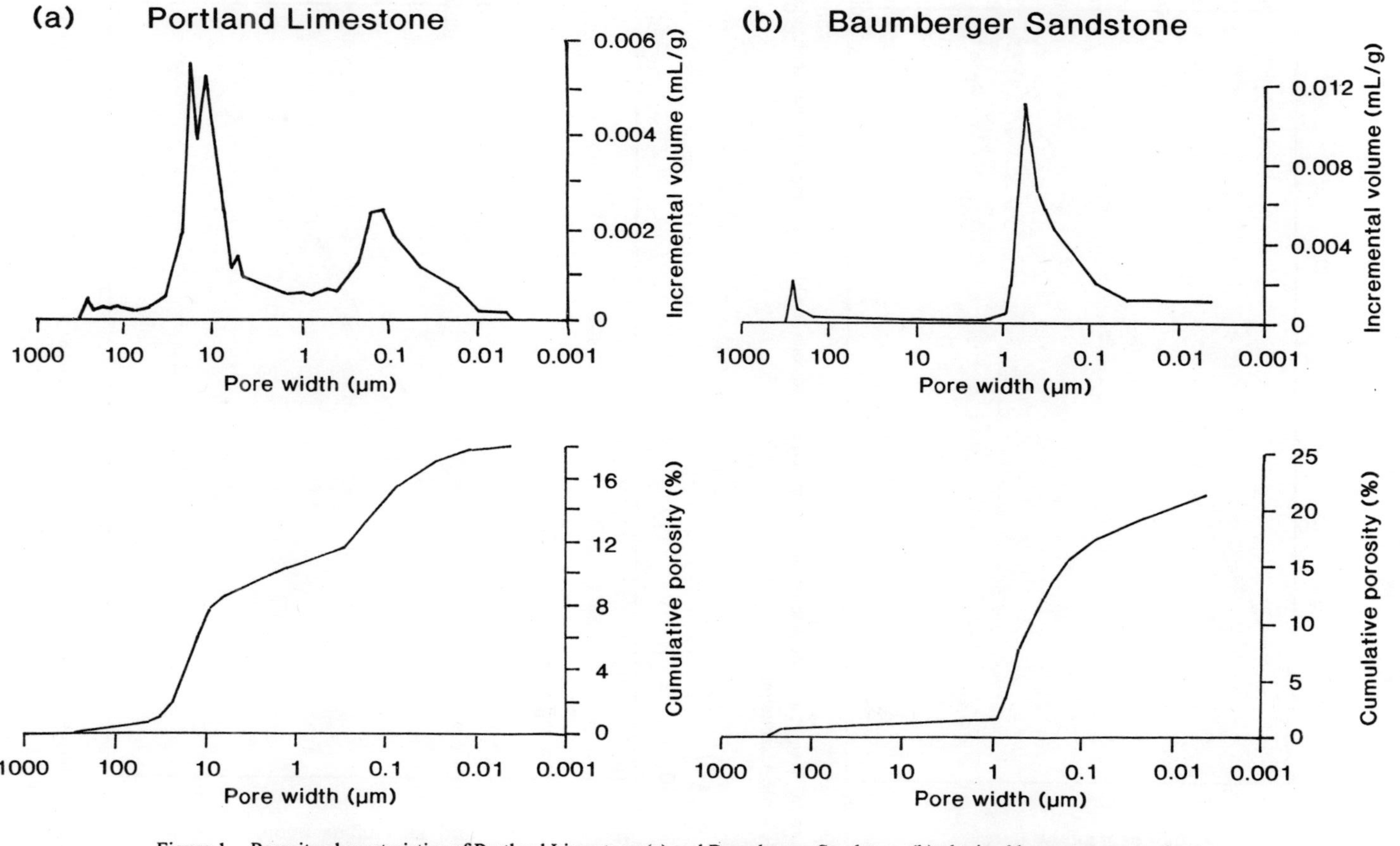

Figure 1. Porosity characteristics of Portland Limestone (a) and Baumberger Sandstone (b) obtained by mercury porosimetry.

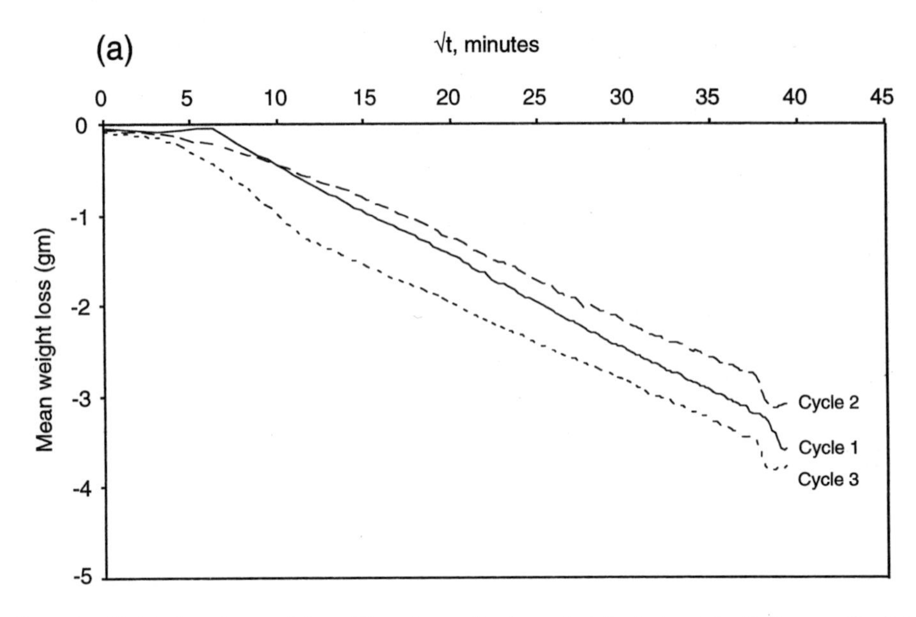

Figure 2a. Weight loss characteristics of Baumberger Sandstone test blocks treated with three applications of 10% $MgSO_4$ solution.

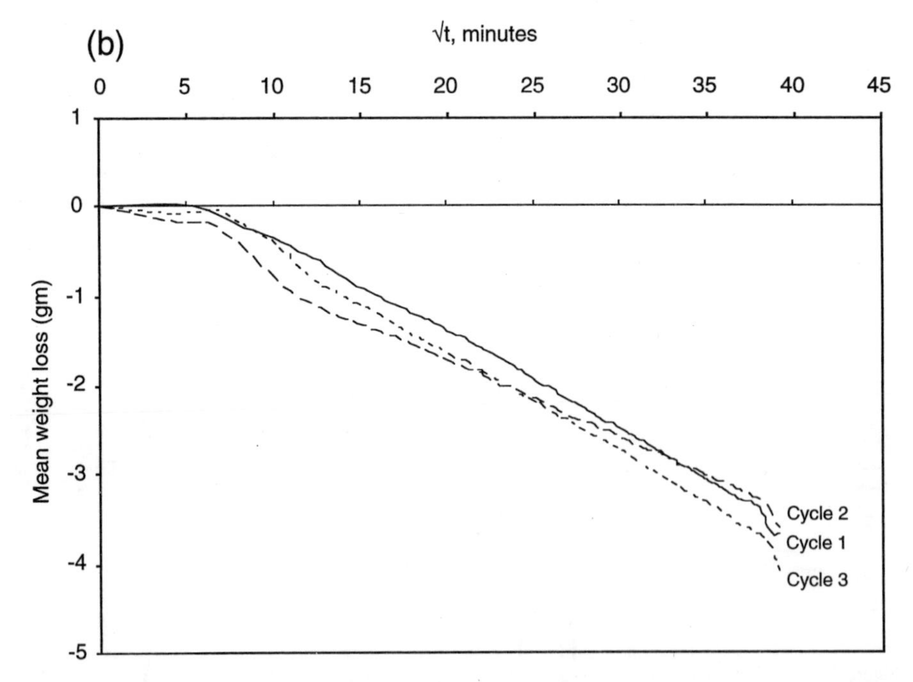

Figure 2b. Weight loss characteristics of Baumberger Sandstone test blocks treated with three applications of 10% Na_2SO_4 solution.

In the case of Portland Limestone (Figures 3a and 3b), relationships between the moisture losses at the end of the different cycles is the same as that for Baumberger Sandstone, in that the third cycles showed the greatest losses and the second cycles the least. However, at the beginning of cycles the patterns of moisture loss are different from those of Baumberger. The moisture loss curves (plotted against the square root of time) are initially convex and indicate that Portland must be placed under a more prolonged and hence greater temperature stress before moisture is lost. This in turn could reflect the higher suction at which moisture is held within the finer micropores of this stone and/or the greater depth to which salt solution had penetrated *via* the larger pores.

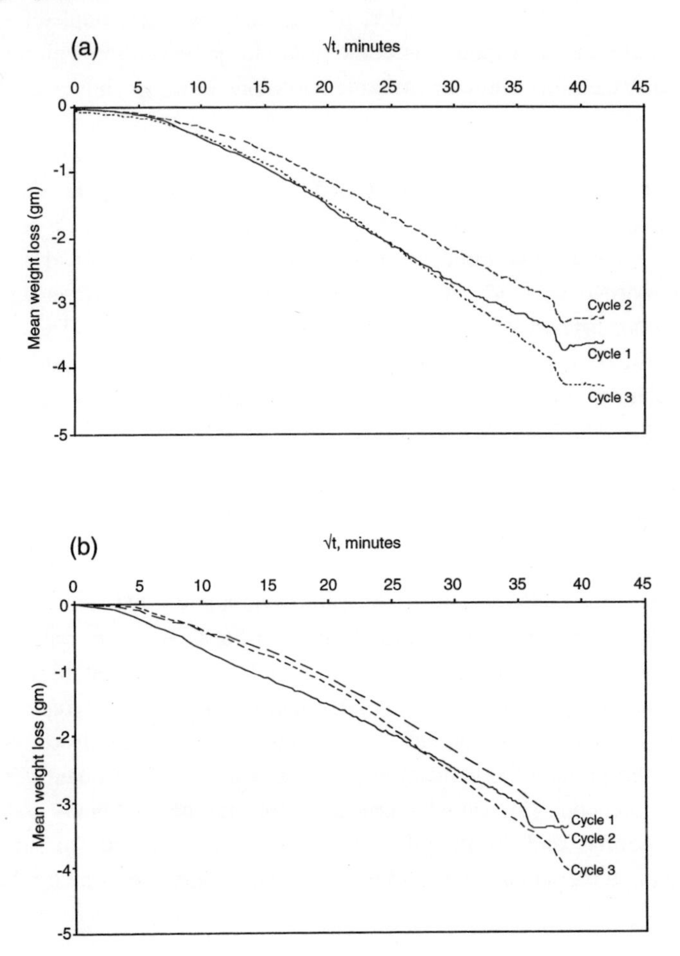

Figure 3a. (above) and b. (below). Weight loss characteristics of Portland Limestone test blocks treated with three applications 10% $MgSO_4$ solution (Figure 3a) and 10% Na_2SO_4 solution (Figure 3b) (Note, cycle 1 was prematurely terminated by a power failure - Figure 3b).

Finally, on all curves there are phases of rapid, short-lived moisture loss at the end of the cycle when air temperature dropped from 38°C to 20°C over 30 minutes. This effect is similar to that observed in winter months over deep water bodies where, in accordance with Dalton's Law, evaporation from the surface is proportional to the difference between the saturation vapour pressure at the surface temperature and the actual vapour pressure of the overlying air [9]. The combination of a low air temperature and residual heat within the blocks would therefore be expected to trigger an increase in the rate of evaporation. In a recently wetted block this might be expected to draw moisture from within the near surface zone and to trigger the crystallisation of a surface efflorescence. However, at the end of a drying cycle it would seem more likely that, provided the surface was not completely sealed, an episode of rapid moisture vapour loss could promote crystallisation within the stone.

The results therefore show identifiable variations in the drying characteristics of the two stone types after only three cycles of a weathering simulation. The differences between cycles are most marked in blocks treated with $MgSO_4$ (Figures 2a and 3a) and might indicate the salt dependency of drying characteristics. In which case the clearer demarcation between values at the end of each cycle could be a product of the lower solubility and more rapid crystallisation of $MgSO_4$ at 38°C [10]. However possibly of more importance in this instance, than absolute differences in drying curves, is the consistency of relationships between the three cycles. This would seem to imply that the observed effects are real and that at least under laboratory conditions the accumulation of salt modifies stone response to heating/cooling and wetting/drying.

4. Conclusions

The consistent variations in patterns of moisture loss suggest that care should be taken in ascribing the results of simulation and durability tests solely to repetitions of the initial test regime. For example, sudden weight losses from test blocks are often attributed to cumulative, low magnitude fatigue effects produced by repeated cycling. An alternative interpretation could be that catastrophic breakdown is triggered primarily by the preferential concentration of salts in sufficient quantities at certain depths within the stone [3]. Thus the crucial factor may be how many and what type of heating/cooling and wetting/drying cycles are required to achieve this concentration in a particular stone, and not how many times the stone has been placed under stress

The sudden increase in evaporation associated with the drop in air temperature at the end of each cycle also requires further investigation. If salt solutions within a stone are near to saturation this evaporation could lead to rapid crystallisation, particularly at or just beneath the stone surface. As such it might provide an additional stimulus to decay triggered by the crossing of regular environmental thresholds including any rapid fall in air temperature associated with, for example, sunset each day.

In light of the preliminary observations described here, further experiments are in progress to investigate the effects of a range of thermal regimes and salt types at different concentrations. In order to explain the observed effects of porosity and salts on moisture loss, future work will also relate patterns of moisture loss to the distribution of salts by depth, assessed by ion chromatography, and the crystallisation of salts within pores, examined by thin section. In this way it is hoped to obtain information that will both assist in the explanation of patterns of decay and inform the design of future simulation and durability tests.

5. References

1. Cook R. U., Laboratory simulation of salt weathering processes in arid environments, *Earth Surface Processes and Landforms* **4** (1979), 347-359.
2. Smith B. J., Weathering processes and forms, in: *Geomorphology of Desert Environments*, ed. Abrahams A. D. and Parsons A. J. (Chapman and Hall, London 1994), 39-63.
3. Smith B. J. and McGreevy J. P., Contour scaling of a sandstone by salt weathering under simulated hot desert conditions, *Earth Surface Processes and Landforms* **13** (1988), 697-706.
4. Snethlage R. and Wendler E., Moisture cycles and sandstone degradation, in: *Saving our Cultural Heritage*, ed. Baer N. S. and Snethlage R. (Wiley, Chichester, 1996), 7-24.
5. Smith B. J. and McAlister J. J., Observations on the occurrence and origins of salt weathering phenomena near Lake Magadi, southern Kenya, *Zeitschrift für Geomorphologie* **30** (1986), 445-460.
6. Smith B. J. and McGreevy J. P., A simulation study of salt weathering in hot deserts, *Geografiska Annaler* **64a** (1983), 127-133.
7. Goudie A. and Viles H., Salt *Weathering Hazards* (1997) Wiley, Chichester.
8. McGreevy J. P. and Smith B. J., Salt weathering in hot deserts: observations on the design of simulation experiments, *Geografiska Annaler* **65a** (1983), 161-170.

9. Ward R. C. and Robinson M., *Principles of Hydrology* (1990) McGraw Hill, London.
10. Sperling C. H. B. and Cooke R. U., *Salt Weathering In Arid Environments. I. Theoretical Considerations* Bedford College, University of London, Papers in Geography, **8** (1980).

6. Acknowledgements

The writers are indebted to Patricia Warke, Paul Carey and John Meneely for their assistance in carrying out the drying experiments, to Susana Pombo of The Robert Gordon University for providing the porosity data and to Gill Alexander for preparing the diagrams.

Stone Weathering and Atmospheric Pollution Network'97: Aspects of Stone Weathering, Decay and Conservation.
Edited by M.S. Jones & R.D. Wakefield © 1998 Imperial College Press.

CHARACTERISATION OF DECAY FEATURES ON SANDSTONE FOLLOWING CLEANING: PRELIMINARY OBSERVATIONS

P.A. WARKE, B.J. SMITH, U.R. CAMPBELL
School of Geosciences, The Queen's University of Belfast,
Belfast, BT7 1NN, Northern Ireland.
Tel. (01232) 273350. Fax (01232) 321280.
E mail p.warke@qub.ac.uk
E mail b.smith@qub.ac.uk

Cleaning of a complex Triassic sandstone building resulted in extensive microbial colonisation and 'greening' of stone surfaces. In addition to its unsightly appearance, SEM examination suggests that biogeochemical and biogeophysical processes may be contributing to weathering and breakdown of the sandstone. This case study highlights the importance of an informed approach to stone cleaning and adoption of a conservative approach in cases where potentially adverse long-term effects may outweigh the short-term benefits of cleaning.

1. Introduction

Soiling of buildings and monuments has occurred as a result of past and present fossil fuel combustion derived from domestic, industrial and vehicular sources. Development of black stains and crusts, while detracting from the aesthetic appeal of a structure, may also contribute to the breakdown of materials through the development of a range of decay features [1]. The decision to clean buildings, however, is generally made on the basis of improving their overall appearance with little thought given to the potential long-term effects on the structural integrity of building materials. Despite the extensive practice of stone cleaning, there is as yet no method that does not cause some degree of damage or alteration to the underlying stone and there are no studies which allow prediction of the long-term structural and visual consequences [2]. Indeed, debate persists regarding the benefits of stone cleaning and it has been suggested that black crusts and hardened, stained surfaces may, in some instances, provide a protective barrier, shielding weakened substrate from the worst effects of sub-aerial weathering processes. If this outer layer of stone is removed during cleaning it may result in accelerated short-term breakdown of stone and initiation of a series of long-term decay processes [3].

To illustrate the possibly deleterious effects of stone cleaning and treatment, this study reports surface alteration and decay features affecting a 125 year old sandstone church following selective cleaning and mortar replacement.

2. Site location and description

Fitzroy Presbyterian Church was built between 1872 and 1874 in the city of Belfast, Northern Ireland. It is located in a built-up area which, despite 'clean air' legislation, remains subject to high levels of atmospheric sulphur and particulates, derived primarily from domestic sources. Much of Belfast lies within the incised valley of the River Lagan bounded by areas of higher ground to the north and south. Because of this location Belfast is particularly prone to temperature inversions under anticyclonic conditions during which pollution is trapped within the valley, particularly in the city centre [4].

The main body of the building is constructed of Scrabo stone, a local quartz sandstone of Triassic (Bunter) Age. Scrabo sandstone is characterised by its structural, mineralogical and textural variability with well defined laminations and bedding planes [5]. The colour of the stone is also extremely variable with cream, grey, buff and pink stone being the most commonly used in buildings. Porosity varies between 14.8%–17.4% [5, 6], because of the differential packing of mineral grains and the presence of concentrations of authigenic clay minerals – in particular smectites. These clay minerals tend to be more abundant in fine-grained laminations and less so where grains are loosely packed [7].

2.1 Refurbishment procedure

By the early 1980's the building exhibited extensive evidence of general soiling and black crust development – primarily the result of over 100 years exposure to emissions from fossil fuel combustion. Because of this, a two-phase refurbishment programme was initiated and funded by the Church authorities in 1987 with original plans envisaging complete cleaning of the building. However, in response to results from the first phase, cleaning was eventually restricted to the north front and tower with mortar replacement on the remaining facades (Figure 1).

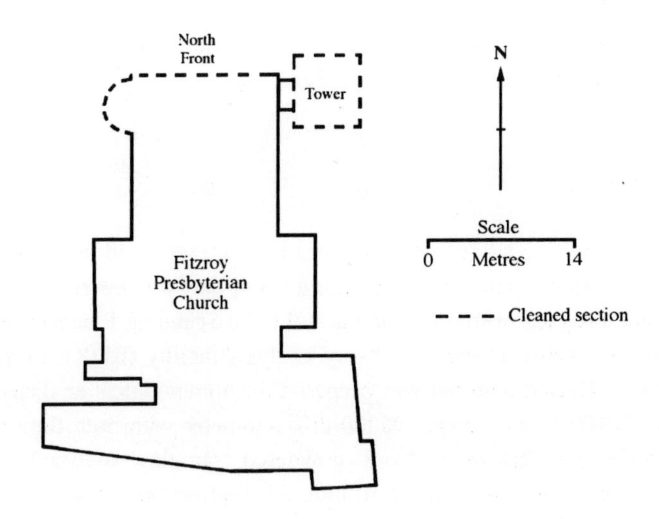

Figure 1. Plan of Fitzroy Presbyterian Church showing section of cleaned stone.

2.2 Cleaning procedure

Information regarding the cleaning procedure was verbally obtained from the architects involved, but was limited in detail especially with regard to the products used.

The front and adjoining tower were treated with an alkaline wash (degreaser) followed by the application of acid cleaning agents (hydrofluoric acid – HF) and then a hot water, low pressure wash off. Stone on the upper section of the tower and spire was also treated with consolidants following cleaning.

Within a year of completing the first phase of cleaning, it became evident that the cleaned stone surfaces were being extensively colonised by algae through a gradual 'greening' of the front facade. Repeated treatment with a herbicidal wash was only effective for short periods with recurrence of algal growth after several months. In some sheltered locations on the cleaned facade lichens, mosses, grasses and ferns have become well established.

2.3 Mortar replacement

On the remaining facades south, west and east facing, cleaning was abandoned in favour of mortar replacement. The old mortar was mechanically removed causing damage in some places to the adjacent stone. The original mortar appears to have been replaced with a modern, cement-based material; however, no information regarding its exact composition was available.

3. Methodology

Small samples of loose stone were collected from cleaned and uncleaned facades, but given the current status of this building, sampling was necessarily restricted. Samples were analysed primarily using a Jeol 6400 Scanning Electron Microscope (SEM) with an Energy Dispersive X-ray analysis facility (EDX). In addition to SEM analysis, selected material was prepared for mineralogical analysis by X-ray Diffraction (XRD) in a Siemens D5000 diffractometer with both the <63µm and <2µm particle size fractions from powdered samples analysed. Elemental components were also analysed by Atomic Absorption Spectroscopy (AAS) and Ion Chromatography (IC).

4 Results and discussion

A visual inspection of the building identified clear differences between stone surface characteristics of the cleaned and uncleaned facades with SEM analysis showing the microscopic components of these surfaces.

4.1 Visual inspection

Facades that had not been cleaned retained a blackened appearance with extensive soiling of stone surfaces and black crust development in sheltered sites. The structural and mineralogical variability of Scrabo sandstone means that not all blocks display a similar extent of surface change or the same decay features. Despite soiling some blocks appeared to retain their structural integrity with original tooling marks still visible on the stone surface while adjacent blocks exhibit extensive evidence of active decay and surface retreat. On some blocks the outer soiled layer of stone had been detached from the underlying material, exposing a friable substrate with evidence of active material loss primarily through granular disintegration and multiple flaking. In the latter, individual flakes were generally less than 2–3mm in thickness and less than 2cm in diameter and occurred most frequently on fine-grained blocks. These decay features were most obvious in the lower 1–1·5m of the walls but were not restricted to this area.

In some stones, decay appears to be a relatively recent phenomenon as indicated by the extent of block retreat relative to the recently repointed mortars which can stand proud of the receding block surfaces by 1–2cm. When repointing it is important to ensure that replacement mortar is lower in strength than adjacent stone, being sufficiently adaptable to resolve stresses arising from repeated expansion and contraction of stone due to heating and cooling and wetting and drying [8]. In this particular example, the replacement mortar would appear to be structurally stronger than the Scrabo sandstone. This has contributed to the accelerated decay of some blocks since repointing, by restricting moisture movement and any expansive stresses initiated within the sandstone.

It is important to note that none of the uncleaned facades exhibited evidence of algal colonisation comparable to that observed on the cleaned front of the building, though algae may well be present but masked by the soiling.

The cleaned facade of the building comprises extensive areas where stone surfaces are green in appearance, most commonly on surfaces over which rainwater flows. Areas devoid of algal colonisation are typically sheltered from both rainwash and surface flow, for example, in doorways and under overhangs. The 'greening' of the facade affects the full vertical extent of the building with the exception of the upper sections of the tower and spire which, coincidentally, were treated with consolidants after cleaning. The application of consolidant may have reduced stone surface porosity and/or created toxic surface conditions which have prevented biological colonisation. The 'greening' of the front facade is unsightly obscuring surface detail and colour of the cleaned sandstone and has prompted repeated attempts, through the application of herbicides, to restore the building to its 'clean' condition.

4.2 SEM examination of cleaned stone

Samples of cleaned stone from green and non-green surfaces were examined using SEM.

Sandstone samples with surface 'greening' showed extensive development of algae cover on surface and within the substrate, with individual algal cells less than 10μm in diameter in their desiccated form (desiccation due to SEM sample preparation, Figures 2a and 2b). Gypsum development was also evident in the substrate as well as the penetration of fungal hyphae from surface to subsurface material. Isolated fly-ash (oil and coal) particles were present on surface samples and were invariably incorporated into the algal cover (Figure 2a).

Evidence from SEM examination of 'greening' samples suggests that algal and fungal growth may be making a direct contribution to the weathering and breakdown of cleaned stone through the disruptive effects of fungal hyphae and algal filaments. Their penetration from surface to subsurface material may lead to a gradual widening of intergranular spaces and other points of weakness through repetitive application of expansive stresses associated with wetting and drying of cells. In addition to physical disruption of stone, secretion of organic acids may contribute to breakdown through the chemical disruption and resultant weakening of constituent minerals [9]. These biogeochemical and biogeophysical processes have been widely invoked as effective mechanisms of rock breakdown [10-16] and through their effects they may also contribute indirectly to stone breakdown by facilitating ingress of moisture and salts in solution to substrate material.

The presence of algal and fungi on stone surfaces may also facilitate the accumulation of windblown and rain-washed particulate material. It is suggested that algal and fungal cover may act " ...as a precursor for later formation of detrimental crusts..." [17 (p.29)]. As microbial colonies develop, their secretions create a 'sticky' biofilm which may initiate conditions of positive feedback whereby, the biofilm facilitates moisture retention, encouraging further microbial development, and increases capture and adhesion of nutritive aerosols and other micro-organisms [17, 18]. On many of the surface samples examined fly ash particles and other detrital material were observed within the microbial cover (e.g. Figure 2a).

Given that extensive microbial colonisation did not appear to affect the stone to the same extent before cleaning or indeed does not presently affect the remaining uncleaned facades, it would seem that the cleaning process has physically and/or chemically altered the stone, creating conditions conducive to microbial growth and development. Unfortunately, insufficient data exist regarding elemental concentrations in cleaned and uncleaned stone to allow meaningful comparison of the two surfaces, but AAS analysis indicates that concentrations of calcium in particular were generally greater on cleaned than on uncleaned surfaces. However, colonisation of the cleaned surface may not be solely the result of chemical changes in micro-environmental conditions but may also reflect physical changes. Removal of soiled material during the cleaning process may expose a weakened substrate and/or open up the stone structure creating conditions conducive to microbial growth with open pore spaces facilitating microbial attachment and the movement of moisture and nutrients.

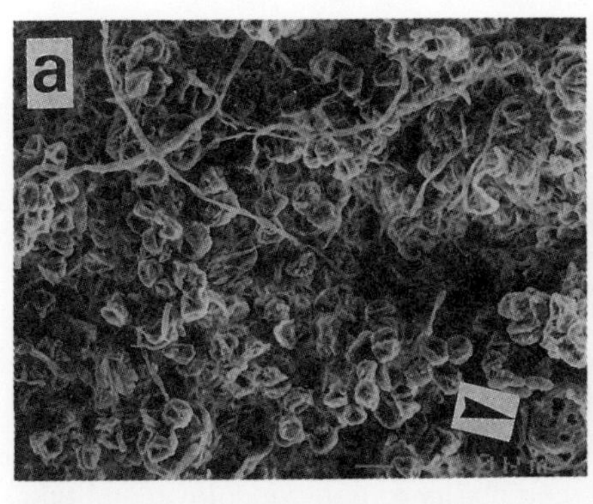

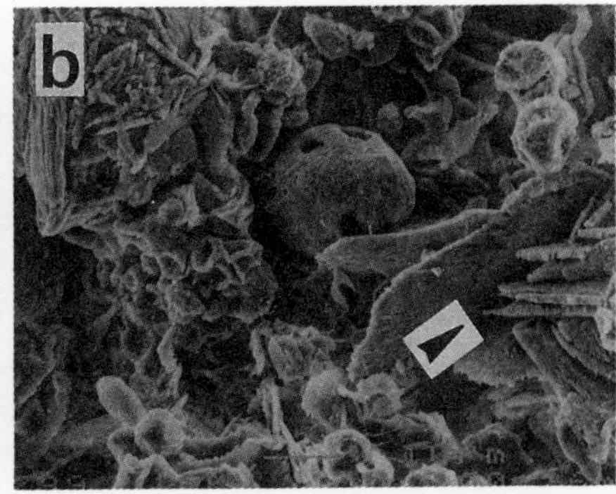

Figure 2. Scanning electron micrographs showing surface and subsurface characteristics of cleaned Scrabo sandstone with extensive surface 'greening'.

(a) Detail of algal cover on surface sample. Individual algal cells are less than 10μm in diameter in their desiccated form. Part of a fine network of fungal hyphae is visible which penetrate into subsurface material but which also help entrap windblown and rain washed detrital material such as the oil fly ash particle shown in this micrograph (arrow) – picture width 150μm.

(b) Detail of subsurface material showing algae and gypsum deposits (arrow) where gypsum occurs as large euhedral crystals or lathes some 15–25μm across. An oil fly-ash particle has been incorporated into the algal cover and was possibly deposited shortly after cleaning (centre of micrograph) – picture width 75μm.

Samples of cleaned stone- 'no greening' from the front facade that showed no evidence of surface 'greening' displayed extensive gypsum development under SEM examination. Closely interlocked gypsum laths with isolated fly-ash particles incorporated into their structure completely covered all surface samples examined (Figures 3a and 3b). The presence of gypsum identified by SEM and EDX was confirmed by XRD analysis with additional evidence provided by AAS and IC, which showed high levels of calcium and sulphate, respectively. This surface cover occurred on cleaned stone sheltered from direct rainwash and differed from the 'green' surfaces through the absence of algae.

Development of gypsum crusts is linked to the absence of rainwash on stone surfaces and hence the removal of precipitated gypsum [19]. The accumulation of gypsum deposits will alter surface porosity characteristics by effectively sealing pore spaces trapping moisture and salts within the substrate. This may have particularly serious implications for stone affected by capillary rise of groundwater. Identification of extensive gypsum deposits on the cleaned facade of Fitzroy Presbyterian Church shows the rapidity of crust development in what is still a 'polluted' environment and serves to highlight the danger of initiating an extensive stone cleaning programme based on the assumption that atmospheric pollution is no longer a problem. Where stone surfaces are exposed to rainwash, gypsum crust development appears to be restricted with conditions at the stone/air interface favouring algal colonisation. Algae appear to rapidly colonise microcracks and crevices on the stone surface and on similar surfaces have been shown to be relatively resistant to detachment by surface water flow [20]. Given the apparent stability of such surfaces and their ability to trap windblown and rain washed particulate material, it is possible that if the concentration of particulates was high enough, black crusts could eventually develop, even in rain washed areas.

The development of extensive algal cover on green samples and a gypsum crust on non-green samples from the cleaned facade suggests that the cleaning process, alkaline degreasing and acid cleaning (HF), altered the chemical and physical characteristics of the Scrabo sandstone surface. Substrate material may also have been affected because chemical cleaning agents can penetrate to a depth of 1cm or more in previously weathered and polluted sandstone leaving a residue which is not removed by subsequent washing [21]. In addition to a chemical cleaning residue, acid-cleaning agents can react with substances within the stone such as insoluble sulphate and constituent mineral grains to form soluble elements, which are retained within the stone fabric.

Cleaning the front facade of Fitzroy Presbyterian Church was of only short-term benefit with regard to the appearance of the building. The procedure, however, appears to have initiated a series of adverse surface changes which will probably require further intervention.

5. Conclusion

Though brief, this case study emphasises the importance of a cautious approach to stone cleaning and highlights the need for a thorough understanding of the material being treated and the characteristics of the building as a whole. Where structurally and mineralogically complex stone is involved it may be necessary to apply several different cleaning procedures and test each first on a small area of stone before treating the whole building. Despite the fact that trial cleaning is widely advocated [22]; this case study demonstrates the damage that can be done when good practice guidelines are not adhered to. This specific case study also provides support for a conservative approach where the potential adverse long-term effects may outweigh the short-term benefits to be gained by cleaning, especially in an environment where levels of atmospheric pollution remain high. As stated by Sasse and Snethlage, "..soiling of the surface should not serve as the sole determinant for cleaning...it must be proved that the deposits are in fact harmful for the stone and that they would cause irreversible damage if they were not removed" [23 (p.233)].

Relationships between cause and effect are not always initially well defined and consequently it is often too easy to end up exchanging one set of problems for another, potentially more serious set, on the basis of ill-informed but well-intentioned decision making.

Figure 3. Scanning electron micrographs showing surface characteristics of cleaned Scrabo sandstone without 'greening'.

(a) Detail of gypsum cover on stone surface. This crust comprises closely interlocked gypsum laths, which completely cover the stone surface – picture width 120μm.

(b) Micrograph showing detail of oil fly-ash particles incorporated into the gypsum crust – picture width 85μm.

6. References

1. Smith B.J. and Magee W.R., Granite weathering in an urban environment: an example from Rio de Janeiro. *Singapore Journal of Tropical Geography*, **2** (1990), 143–153.
2. Maxwell I., Stonecleaning – for better or worse? An overview. in *Stone cleaning and the nature, soiling and decay mechanisms of stone*, ed. Webster R.G.M. (Donhead Publishing, London, 1992), 3–49
3. Smith B.J., Scale problems in the interpretation of urban stone decay, in *Processes of Urban Stone Decay*, ed. Smith B.J. and Warke P.A. (Donhead Publishing, London) 1996, 3–18.
4. Smith B.J. Whalley W.B. and Magee R.W., Background and local contributions to acidic deposition and their relative impact on building stone decay: a case study of Northern Ireland, in *Acid Deposition: Origins, Impacts and Abatement Strategies*, ed. Longhurst J.W.S. (Springer-Verlag, Berlin) 1991, 241–266.
5. Smith B.J. and McGreevy J.P., A simulation study of salt weathering in hot deserts. *Geografiska Annaler*, **65A** 1–2 (1983), 127–133.
6. McGreevy J.P., Some field and laboratory investigations of rock weathering, with particular reference to frost shattering and salt weathering. (Unpublished PhD. Thesis, The Queen's University of Belfast) 1982.
7. McGreevy J.P. and Smith B.J., The possible role of clay minerals in salt weathering. *Catena*, **11** (1984) 169–175.
8. Ashurst J. and Ashurst N., *Practical building conservation, Stone masonry.* (Gower, Aldershot) 1989, 1.
9. Krumbein W.E., Microbial interactions with mineral materials. in *Biodeterioration* ed. Houghton D.R. Smith R.N. and Eggins H.O.W. (Elsevier, London) 1988, 7.
10. Lewis F. May E. and Bravery A.F., Isolation and enumeration of autotrophic and heterotrophic bacteria from decayed stone, in *Proceedings of the 5th International Conference on Deterioration and Conservation of Stone*, ed. Felix G. (Lausanne, Switzerland, Presses Polytechniques Romandes.) **2** (1985) 633–641.
11. Seaward, M.R.D., Lichen damage to ancient monuments: a case study. *Lichenologist*, **20** (1988), 291–295.
12. Cooks J. and Otto E., The weathering effects of the lichen *Licidea* Aff. *Sarcogynoides* (Koerb.) on Magaliesberg Quartzite. *Earth Surface Processes and Landforms*, **15** (1990), 491–500.

13. Hall K. and Otte W., A note on biological weathering on nunataks of the Juneau Icefield, Alaska. *Permafrost and Periglacial Processes*, **1** (1990), 189–196.
14. Moses C.A. and Smith B.J., A note on the role of the lichen *Collema auriforme* in solution basin development on a Carboniferous limestone substrate, *Earth Surface Processes and Landforms*, **18** (1993), 363–368.
15. Koestler R.J. Brimblecombe P. Camuffo D., How do environmental factors accelerate change? In *Dahlem Workshop Report ES **15** Durability and change: the science, responsibility, and cost of sustaining cultural heritage*, ed. Krumbein W.E. Brimblecombe P. Cosgrove D.E. and Staniforth S., (J.Wiley, Chichester), 1994, 149–163.
16. McCarroll D. and Viles H., Rock weathering by the lichen *Lecidea auriculata* in an arctic alpine environment. *Earth Surface Processes and Landforms*, **20** (1995), 199–206.
17. Koestler R.J. Warscheid T. and Nieto F., Biodeterioration: risk factors and their management. in *Dahlem Workshop Report Saving our architectural heritage, the conservation of historic stone structures*, ed. Baer N.S and Snethlage R. (J.Wiley, Chichester) 1997, 25-36.
18. Characklis W.G. and Wilderer P.A., Structure and function of biofilms, in *Dahlem Workshop Report, Structure and function of biofilms*, ed. Characklis W.G. and Wilderer P.A. (J. Wiley, Chichester) 1989, 5-18.
19. Whalley W.B. Smith B.J. and Magee R.W., Effects of particulate air pollutants on materials: investigation of surface crust formation, in *Stone cleaning and the nature, soiling and decay mechanisms of stone*, ed. Webster R.G.M. (Donhead Publishing, London) 1992, 227–234.
20. Rittman B.E., Detachment from biofilms. In, *Dahlem Workshop Report, Structure and function of biofilms*, ed. Characklis W.G. and Wilderer P.A. (J. Wiley, Chichester 1989), 49–58.
21. MacDonald J. Thomson B. and Tonge K., Chemical cleaning of sandstone–comparative laboratory studies, in *Stone cleaning and the nature, soiling and decay mechanisms of stone*, ed. Webster R.G.M., (Donhead Publishing, London 1992), 217–226.
22. Ashurst N., *Cleaning Historic Buildings*, Volumes 1 & 2 (Donhead Publishing, London, 1994)
23. Sasse H.R. and Snethlage R., Methods for the evaluation of stone conservation treatments, in *Dahlem Workshop Report Saving our architectural heritage, the conservation of historic stone structures*, ed. Baer N.S. and Snethlage R. (J.Wiley, Chichester, 1997), 223–243.

7. Acknowledgements

The authors would like to thank the authorities at Fitzroy Presbyterian Church for allowing access to the building, Leighton Johnston Associates (Architects) for information about the cleaning procedure and staff in the SEM unit at Queen's University for their assistance.

Stone Weathering and Atmospheric Pollution Network '97: Aspects of Stone Weathering, Decay and Conservation.
Edited by M.S. Jones & R.D. Wakefield

REMOVAL AND ANALYSIS OF SOLUBLE SALTS FROM CHEMICALLY CLEANED SANDSTONES

S. POMBO FERNANDEZ[†], K. NICHOLSON*[†] D. URQUHART[†]
[†]*Masonry Conservation Research Group*
**Environmental Geochemistry Research Group*
The Robert Gordon University, Aberdeen, AB25 1HG, Scotland
E mail s.pombo@rgu.ac.uk

The concentration of soluble ions (Mg, Na, K, Cl, NO_3, SO_4 and F) leached from two different sandstones, identified as blonde Crossland Hill and red Locharbriggs, immersed in deionised water were monitored over a period of 30 days. This desalination method is commonly used in conservation and was modified for the removal of salts from chemically cleaned sandstones. Results are reported on the use of supressed ion chromatography, atomic absorption and fluoride electrode as methods for qualitative and quantitative determination of ions leached from sandstones during desalination procedures. In all tests, a slow initial salt-removal process occurred. The desalination rate increased until after approximately two weeks where the total amount of salt removed remained contant, although the process was not yet completed even after 30 days (e.g. Mg, 800 ppm/mgL^{-1}; Na, 7000 ppm/mgL^{-1} and K, 1500 ppm/mgL^{-1} from cleaned Locharbriggs sandstone). The pH of the leached solution was monitored to assess the buffering capacities of the stone and the changes on the proportion of carbonate.

1. Introduction

Salt has been associated with a variety of decay features on sandstone buildings such as spalling, blistering, contour scaling, efflorescence and fracturing. The decay features have all been attributed to various mechanistic processes involving temperature and moisture cycling, salt crystallisation and salt migration [1-4].

The principal cause of stone decay in urban environments in the United Kingdom is the crystallisation of salts within the pores of stone [5-8]. Resistance to crystallisation damage is strongly dependent on the internal fabric of the stone and decreases as the proportion of fine pores increases. Crystallisation caused by freely soluble salts such as sodium chloride (NaCl), sodium sulphate (Na_2SO_4) or sodium hydroxide (NaOH) results in crumbling or powdering of the stone surface. The deterioration caused by soluble salts that can crystallise both as hydrate or anhydrate has been investigated by many authors [9-11]. The general conclusion is that salt decay only happens when a high moisture content and a mass transfer process permits solution, migration and crystallisation of soluble salts. Sodium sulphate, which can crystallise as decahydrate mirabilite ($Na_2SO_4.10H_2O$) or as anhydrous thernadite (Na_2SO_4), is known to cause greater deterioration, and at much faster rates, than a non-hydrating salt such as NaCl. However, it is still not clear whether the increased damage that this salt produces is due to the crystallisation process or to the hydration of the salt.

The present models explaining the causes of salt damage based on the application of Correns and Steinborn's formula [12] have been questioned by Snethlage and Wendler [13]. These authors propose a new model in which dilatation and contraction of salts under the influence of moisture and dissolved ionic species is the only process needed to explain the deterioration of stone. By using this model they relate zone of salt precipitation to the zone of maximum moisture penetration.

In addition to salts derived from the environment and those present within the stone fabric, salts are also introduced by poor cleaning methods, especially chemical cleaning. Chemical cleaning to some extent works by the "creation" of soluble salts which, it is widely assumed, are then dissolved and removed from the stone by rinsing with water. Nevertheless, the presence of efflorescence on a number of chemically cleaned buildings casts some doubt on the validity of this hypothesis [14]. The mechanisms of how chemical cleaning processes may damage the stone are not yet clearly understood, but it is apparent that salt induced decay on buildings can be accelerated after such cleaning treatments.

To understand the problem of sandstone deterioration created by soluble salts one must recognise the sources of these salts and how they bring about their damaging effects. It is assumed that the removal of ions from minerals in the stone matrix is the basic mechanism of chemical weathering, with water being the most important agent. By measuring the concentration of ions leached from a stone sample during its immersion in water, any trend in the leaching rate could be related to the weathering of sandstone minerals, and also the impact of a cleaning process. Rates of weathering may potentially be enhanced or reduced by a chemical cleaning process, stone porosity and mineralogy of the stone.

2. Experimental

Two sandstone types were used in the studies, Locharbriggs sandstone and Crossland Hill sandstone. These types were selected for the study because of their differences in porosity and mineralogy. Locharbriggs is a red, medium to coarse grained ferruginous quartz arenite Scottish sandstone of Permian age. The mineralogy of Locharbriggs as observed by a cross polarising microscope is: quartz, 91%; microcline feldspar, 2%;

plagioclase feldspar, 2%; orthoclase feldspar, 3%; muscovite mica, 1%. The rest of the rock is made up of opaques and associated clays. The grains are well sorted with high sphericity. The porosity is reasonably high at 22.4% with a saturation coefficient of 0.64. Porosity is intergranular rather than within grains however the feldspar appears porous due to dissolution. There is some alignment of grains to form fine banding. There is some evidence of compaction and pressure solution with overgrowths on the quartz and some feldspar grains. Crossland Hill sandstone is a quartz arenite, light brown fine to medium grained stone of Carboniferous age. The mineralogy as observed by a cross polarising microscope is: quartz, 77%; microcline feldspar, trace; plagioclase feldspar, 2%; orthoclase feldspar, 10%; muscovite mica, 3%; biotite mica, 3%; iron oxide/hydroxide, 5%. It is very well cemented by quartz overgrowth and pressure solution, and within domains of the rock intergranular porosity is effectively zero. Its grain size is slightly coarser, ranging up to 0.35mm. This sample is also likely to be durable building stone with a low permeability to water.

An acid-based cleaning method which has been widely used on Scottish buildings was selected to represent the cleaning treatment in the investigation. The cleaning process was carried out according to manufacturers recommendations

Stone surfaces were pre-wetted with deionised water. Afterwards, a thin coating of alkaline degreaser, NaOH (1:3 diluted) was applied using a brush. After 30 minutes, the samples were rinsed with deionised water (600 psi/4.1 kPa) and then the cleaning solution, hydrofluoric acid (HF) (1:3 diluted) was applied using a brush. After 5 minutes, low pressure water rinsing was used to remove the initial acid residue. The treated area was then rinsed thoroughly a second time using high pressure spray (600 psi/4.1 kPa) until neutral pH is reached on the stone surface.

The measurement of soluble ions in immersion water can give information regarding the weathering relative rates of stone in an extreme situation where samples are continuously irrigated with salt solution. To obtain information regarding the loss of soluble ions from stone under such conditions, an investigation was carried out over a 30 day period to determine the leaching trends of soluble salts from two different sandstone types which had undergone a chemical cleaning regime. To this end, a leaching method was adopted and modified from a desalination study commonly used in conservation [15].

Blocks of sandstone, 80x80x40mm in size were prepared of each stone type. Two replicate samples of each sandstone type were used, one chemically cleaned before immersion and the second untreated and used as a control. Samples were then immersed for 30 days in 3000mL of deionised water in closed plastic containers. Mechanical stirring was applied continuously. Finally, at the end of the sampling period, the samples were reweighed after drying to constant weight at 20°C and 50% relative humidity. Samples from the leachates were collected and analysed by atomic absorption and ion chromatography at different time intervals (0, 1, 3, 5, 10, 15 and 30 days) for the presence of soluble Mg, Na, K, Cl, NO_3, SO_4 and F.

The changes in colour components L* (brightness), a* (green to red) and b* (blue to yellow) as a result of cleaning and immersion were measured on dry samples by using a Minolta Colour Meter CR-200b.

3. Results and Discussion

Data obtained from analysis of leachates were normalised, to enable comparison, by dividing them by the factor: m/V, where V is the volume of immersion water (in litres) and m is the weight (in grams) of the sandstone block. For concentration of ions in ppm, these normalised values gave the ion concentration (ppm) per gram of sandstone immersed in one litre of water. The results from atomic absorption showed a rapid increase in the concentrations of Mg, Na and K for both Blonde and Red sandstone after chemical cleaning during the first ten days, with the establishment of a constant rate after two weeks for the Blonde type, and a trend of increasing concentration for the Red sandstone (Figure 1a-c).

Figure 2a-c presents results for the anions chloride (Cl^-), nitrate (NO_3^{2-}) and sulphate (SO_4^{2-}) by supressed ion chromatography. Results for chloride showed how chemical cleaning of Blonde sandstone appeared to inhibit further leaching of chloride into solution after ten days. Results for NO_3^{2-} showed the same inhibitory trend in the cleaned Red sandstone. Sulphate released from chemical cleaned sandstone appeared at higher concentrations than the control samples in both cases and approximately the same or higher concentrations than those for Cl^- and NO_3^{2-}; the time evolution showed a trend to stabilisation after 15 days.

The sulphate content of the sandstone is an important factor, especially in aged sandstones, such as used in this study, which have been exposed to the urban environment. Sulphate present in less soluble forms, such as $CaSO_4$, may become mobilised by conversion to soluble Na_2SO_4 or H_2SO_4 by treatment with alkalis or acids respectively. Werner [16] supports the idea that the origin of SO_4^{2-} is mainly from atmospheric pollution and that its distribution in soluble form is affected by the cleaning process. The evaluation of retained SO_4^{2-} as well as the SO_4^{2-} leached into water would help to deal with the "memory effect", which is a consequence of retained SO_4^{2-} from earlier exposure to high atmospheric SO_2 concentrations.

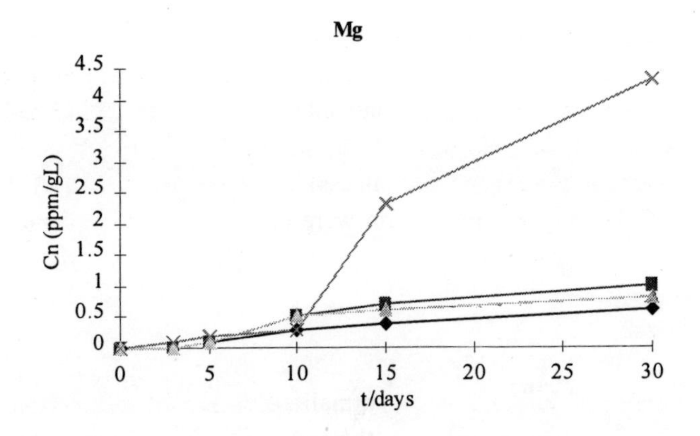

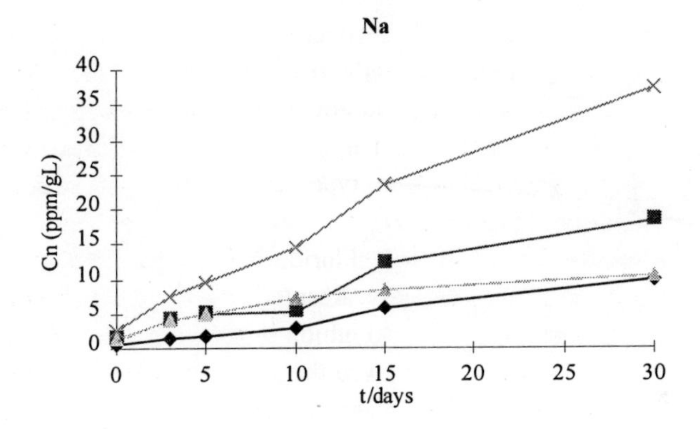

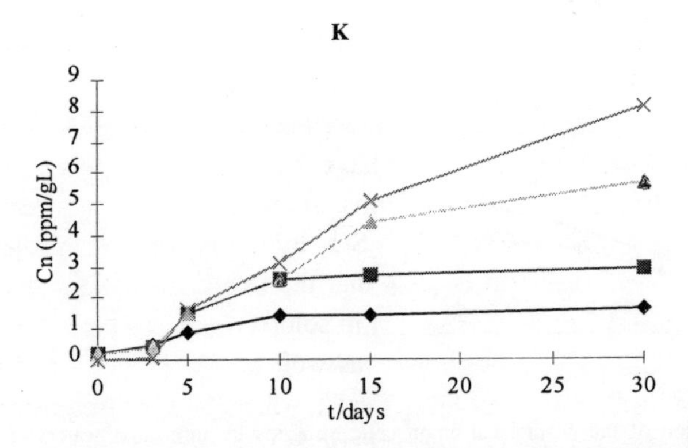

Figure 1a-c. Time evolution of the normalised cation concentrations in immersion waters of sandstone blocks: Blonde control [◆], Blonde cleaned [■], Red control [Δ], Red cleaned [X] (Cn=concentration).

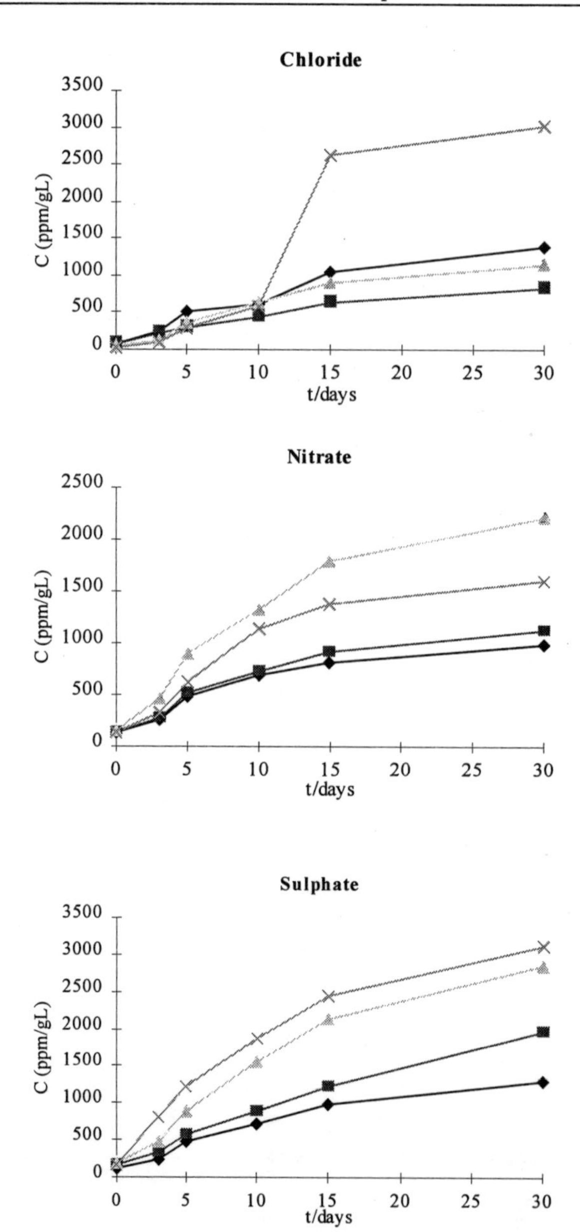

Figure 2a-c. Time evolution of the normalised anion concentrations in immersion waters of sandstone blocks: Blonde control [◆], Blonde cleaned [■], Red control [Δ], Red cleaned [X] (C=concentration).

Figure 3 shows the results for fluoride (F^-) determined by ion selective electrode. As expected, F^- concentrations were higher for both chemically cleaned sandstone samples, increasing rapidly during the first two weeks before stabilised. Fluoride levels for the cleaned Blonde sandstone sample were about one-third of those for the cleaned Red sandstone. This could be explained by the difference in porosity (Blonde, 9.7%; Red, 22.4%) as this would affect the way chemicals are retained in the stone. The fact that the levels of F^- are relatively high after 30 days (1500 ppm/mgL^{-1} for the Red sandstone and 600 ppm/mgL^{-1} for the Blonde sandstone) demonstrates how F^- is readily adsorbed by cations at the mineral surface during the cleaning process, and is not removed totally by rinsing with water. Also, F^- could be precipitated as a metal salt. Only when the stones are immersed in water, as in this experiment, is F^- slowly released from the mineral surfaces in the pore system.

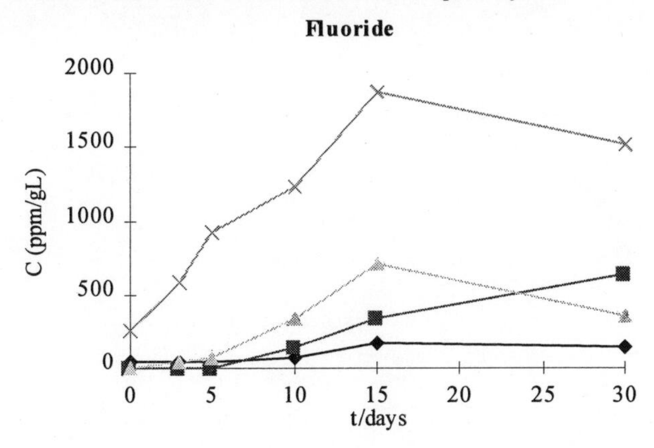

Figure 3. Time evolution of the normalised Fluoride concentration in immersion waters of sandstone blocks: Blonde control [◆], Blonde cleaned [■], Red control [Δ], Red cleaned [X].

The pH of leachate solutions was monitored to show immersion trends during the 30 days experiment, as well as to calculate carbonate (CO_3^-) levels in solution. It is only at pH values close to 6.4 that both species H_2CO_3 and HCO_3^- are present at comparable activities (and hence concentrations). Below pH 6, essentially all the dissolved CO_3^- species are in the form of H_2CO_3, and above pH 7 essentially all are in the form of HCO_3^-. Figure 4 shows the results of pH and log [HCO_3^-/H_2CO_3] in solution. From the evolution of CO_3^- in immersion waters it seems as if the origin of these anions is more related to the experimental approach taken in the study and a reflection of atmospheric CO_2 in equilibrium with $CO_3^=$ in the stone; pH seems not to influence salt leaching.

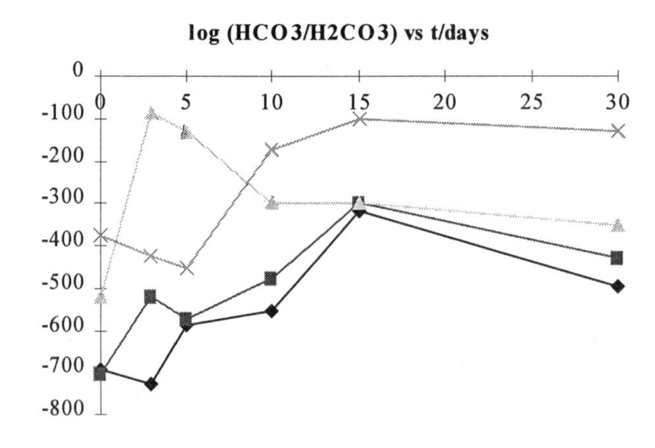

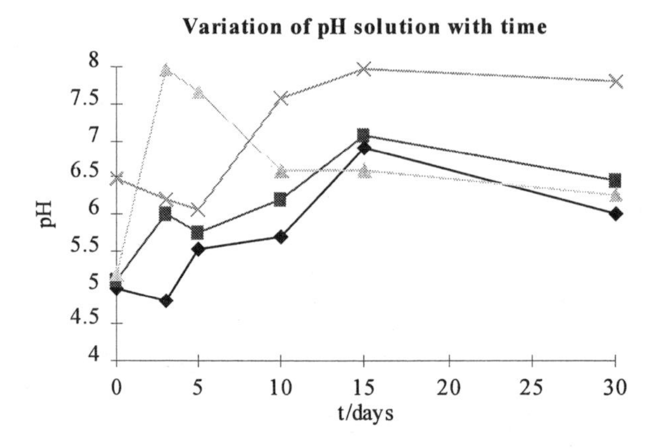

Figure 4. Time evolution of the normalised HCO_3^-/H_2CO_3 concentration and the pH of the immersion waters of sandstone blocks: Blonde control [◆], Blonde cleaned [■], Red control [Δ], Red cleaned [X].

The system developed in this study made use of a sealed bottle partially filled with water, and sandstone blocks immersed in it. In this case, as Livingston [17], calcite solubility in carbonate stone was clearly demonstrated. The supply of CO_2 in the air space of a bottle is limited and therefore the amount absorbed or released during the precipitation or dissolution of calcite will influence its partial pressure.

The rate of release of ions from sandstone blocks into immersion water will be retarded if ions accumulate in the wash solution near the stone surface. At the beginning of the experiment described in this paper, sandstone blocks were immersed in deionised water. For a case such as this, Snethlage and Wendler [13] describe hydric swelling followed by expansion of the stone. As the ions are released into the immersion water the initial salt-free system becomes salt-contaminated, and in this case the hydric dilatation is different; the stone contracts during wetting and expands during drying. The contraction is explained by the formation of dense hydration shells between the stone mineral grains, which become more dense as electrolyte concentration increases. Continuous mechanical agitation of the immersion water would assist in ion removal, but this could also lead to erosion of the stone surface.

The different proportions of clays in both sandstones could help explain the different trends in the release of ions based on cation exchange capacity [18]. Red sandstone contains a larger proportion of clays such as vermiculite, illite and kaolinite, which are capable of adsorbing cations, while the Blonde type only contains traces of kaolinite.

In both sandstones, cleaning had an effect in chromaticity (Table 1); replicate readings of L* showed values increased with cleaning indicating an increase in brightness while a* decreased which implied a shift towards green (or a less red colouration). The b* values decreased slightly in the positive range and therefore yellow dominated. Comparison of surface colour after immersion in water solution showed a similar trend for the Red sandstone. The a* value increased for the Blonde cleaned sandstone after immersion, due to reddening of the surface.

Table 1. Variation of colour after leaching for 30 days (rsd ± 0.5).

	Initial colour			After 30 days		
	mean L*	mean a*	mean b*	mean L*	mean a*	mean b*
Red control	67.16	7.4	13.43	66.86	7.26	13.3
Red cleaned	68.43	4.46	11.86	68.56	4.06	11.13
Blonde control	69.4	1.76	14.46	71.13	0.73	13.03
Blonde cleaned	73.3	0.1	13.03	69.03	2	13.26

Changes in L* could be explained by the appearance of white efflorescences. These efflorescences, not easily identified under SEM, were analysed for presence of cations and anions by atomic absorption spectroscopy and ion chromatography (Figure 5). The results showed large amounts of Cl^- (mainly as KCl and NaCl) as well as SO_4^{2-} which could be responsible for the white patches observed by eye at the stone surface. The observation of an ochre-brown coloration of the Red cleaned sandstone solution also indicated the possible presence of iron leached from biotite and mica and oxidised into the insoluble ferric hydroxide.

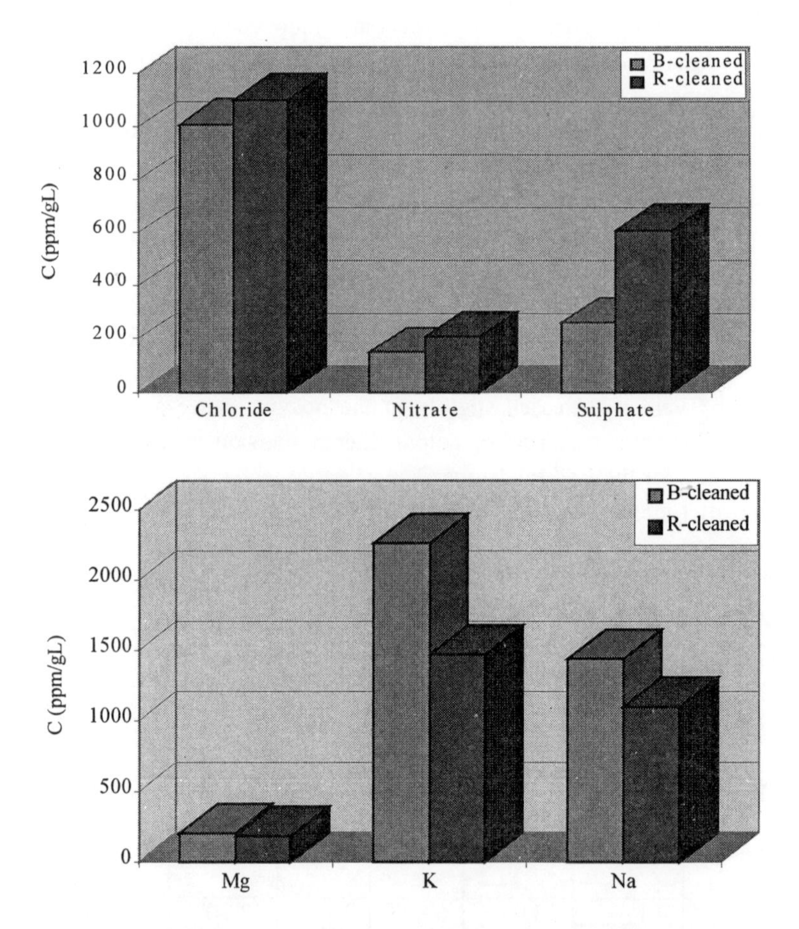

Figure 5. Distribution of soluble anions and cations on the surface of the Red and blonde cleaned sandstone.

4. Conclusions

The differences in the leaching distribution of soluble ions from Red and Blonde sandstone can be explained by different mineralogy since the Red sandstone contains clays such as kaolinite, vermiculatite and illite whereas the Blonde sandstone has only trace kaolinite. Further quantification of these clays will help to explain the possible contribution to the leaching trends for the different soluble ions. Hydrofluoric acid hydrolyses clays and releases Al as well as other elements such as Mg, K, Na, Cl, NO_3, SO_4 and F into solution. Once the distribution of these soluble ions was identified then the:

- distribution in cleaned samples experiencing different environments
- distribution after artificial weathering in a climatic chamber
- distribution after application of different cleaning systems (chemical, abrasive and laser cleaning)

could all be determined.

Before the surface of the sandstone samples were cleaned the source of soluble salts would be expected to be the stone constituents themselves that have migrated from the inside the stone, to the surface. During cleaning, minerals, especially clays, may interact with chemicals in a number of ways; by ion exchange on external surfaces, or adsorption on internal surfaces of crystalline clays. These interactions are pH dependent and competitive, for example, release of exchangeable bases such as Mg, K and Na will contribute to an increase in pH. On the other hand, the solvent action of water on the minerals will substitute bases by solution and leaching. The acid not only supplies H^+ for adsorption but also replaces bases and encourages their solution from the minerals, decreasing pH. Leaching also encourages acidity by removing those metallic cations which might compete with H^+ on the exchange complex.

The movement of ions is only possible by diffusion in water. If pores are not interconnected or have very narrow pore throats, water will not penetrate and any soluble salt will tend to remain on or near the stone surface. The analysis of soluble ions from the surface of both cleaned sandstone types showed comparable levels for Cl^- (1000 ppm/mgL^{-1}) and NO_3^{2-} (< 200 ppm/mgL^{-1}) but SO_4^{2-} was three times higher on the Red stone (600 ppm/mgL^{-1}). Results for cation concentrations showed similar levels of Mg (200 ppm/mgL^{-1}) in both sandstone types; K and Na concentrations were higher in the Blonde stone. Weathering will become damaging at different depths in the stone profile dependent on the initial porosity of the stone. For Red sandstone with a porosity of 24% and 1.9 gcm^{-3} bulk density, it would be expected that more extensive weathering would occur after the 30 days immersion study compared to the more dense Blonde sandstone which had a 9.7 % porosity and 2.4 gcm^{-3} bulk density.

5. References

1. McGreevy J.P., A preliminary scanning electron microscope study of honeycomb weathering of sandstone in a coastal environment, *Earth Surface Processes and Landforms*, **10**, (1985) 509-578.
2. Arnold A. and Zhender K., Salt weathering on monuments. In *The Conservation of Monuments in the Mediterranean Basin*. ed, Zezza F., (Grafo, Bari, 1990), 31-58.
3. Puhringer J. Makes F. and Weber J., Deterioration of Gotland sandstone at the Royal Palace, Stockholm. In *7th International Congress on the Deterioration and Conservation of Stone*. ed. Rodrigues J Henriques F. and Jeremias F (Lisbon, Portugal, 15-18 June, 1992), 687-695.
4. Winkler E.M., Stone in Architecture: Properties, Durability, (Springer-Verlag, Berlin. 3rd edn, 1994).
5. Lewin S.Z., The mechanism of masonry decay through crystallisation, in *Conservation of Historic Stone Buildings and Monuments*, (National Academic Press, Washington, 1982), 120-144.
6. Sperling C.H.B. and Cooke R.N., Laboratory simulation of rock weathering by salt crystallisation and hydration processes in hot arid environments, *Earth Surface Processes and Landforms,* **10**, (1985), 541-555.
7. Goudie A.S., Laboratory simulation of the wick effect in salt weathering of rock, *Earth Surface Processes and Landforms*, **11**, (1986), 275-285.
8. Bell F.G., The durability of some sandstones used in the United Kingdom as building stone, with a note on their preservation. In: *7th International Congress on the Deterioration and Conservation of Stone,* ed. Rodrigues J., Henriques F. and Jeremias F. (Lisbon, Portugal, 15-18 June, 1992), 119-128.
9. Evans I.S. Salt crystallisation and weathering: a review, *Revue de Geomorphologie Dynamique*, **19(4)**, (1970), 153-177.
10. Rossi-Manaresi R and Tuci A., Pore structure and the disruptive or cementing effect of salt crystallisation in various types of stone, *Studies in Conservation*, **36**, (1991) 53-58.
11. Charola A.E. and Weber J., The hydration-dehydration mechanism of sodium sulphate. In *7th International Congress on the Deterioration and Conservation of Stone*, ed. Rodrigues J. Henriques F. and Jeremias F., (Lisbon, Portugal, 15-18 June 1992), 119-128.
12. Correns C.W., Uber die Erklarung de sogenannten Kristallisationskraft, Sitzungsberichte der Preussischen Akamie der Wissenschaften, **11** (1926), 81-88.
13. Snethlage R Wendler E., Moisture Cycles and Sandstone Degradation. Saving Our Architectural Heritage: The Conservation of Historic Stone Structures ed. Baer N.S. and Snethlage R. (1997.)

14. Webster RG.M. ed. Stonecleaning and the nature, soiling and decay mechanisms of stone, Proceedings of the International Conference, Edinburgh, UK. (1992).
15. Costa Pessoa J. Farinha Antunes J.L. Figueiredo M.O. and Fortes M.A., Removal and analysis of soluble salts from ancient tiles, *Studies in Conservation,* **41**, (1996), 153-160.
16. Werner M., Changes to surface characteristics of sandstone caused by cleaning methods applied to historic stone monuments. In Vth Int. Conf. on Durability of Building Materials and Components. ed. Baker J.M. Nixon P.J. Majumdar A.J. Davies H., (Brighton, 7-9 Nov. 1990).
17. Livingston R., Geochemical considerations in the cleaning of carbonate stone Ed. R.G.M. Webster, (Donhead Publishers., London, 1992), 166-179.
18. Wilson M.J., Clays: Their significance, properties, origins and uses, in *A Handbook of Determinative Methods in Clay Mineralogy*.(Chapman & Hall, New York, 1987) 1-25.

6. Acknowledgements

I would like to thank Historic Scotland for their help in the identification of the sandstones used in this study and The Robert Gordon University for funding this PhD project. Also, many thanks to the technical staff from the School of Applied Sciences and School of Construction Property and Surveying for their help and advice and finally to Eileen Kenedy, Queens University Belfast for carrying out clay analysis.

Stone Weathering and Atmospheric Pollution Network'97: Aspects of Stone Weathering, Decay and Conservation.
Edited by M.S. Jones & R.D. Wakefield © 1998 Imperial College Press.

ASPECTS OF MICROCLIMATE: EFFECTS ON THE DISTRIBUTION OF A BIOLOGICAL COMMUNITY ON HERMITAGE CASTLE, UK

M.S. JONES
Roslyn Associates, 19, High Street, Kemnay, Aberdeenshire
Scotland, AB51 5NB.
Tel ++ 44 1467 642114
E mail mzj2@tutor.open.ac.uk

R.D. WAKEFIELD, G. FORSYTH
School of Applied Sciences, The Robert Gordon University
Blackfriars Street Building, Schoolhill, Aberdeen, AB10 1FR
Tel ++ 44 1224 262845 Fax ++ 44 2114 262828
E mail: r.wakefield @rgu.ac.uk

P.J. MARTIN
School of Construction, Property and Surveying, The Robert Gordon University,
Garthdee Road, Aberdeen, Scotland.
Tel: ++ 44 1224 263531 Fax: ++ 44 1224 263777.
E mail p.martin@rgu.ac.uk

Decayed sandstone colonised predominantly by *T.aurea* was found on a 13th Century sandstone castle in the Scottish Borders. Microclimatic monitoring (stone surface temperature, air temperature, radiant heat, relative humidity and light intensity) recorded the environmental conditions experienced by the microbial community during a calendar year (June 95-96). The extreme stone surface temperature recorded on the N face was 28.5°C (max.) -10.50°C (min.) and on the S face, 37.5°C/-8.0°C. Dew-point calculations indicated that condensation occurred for 11% of the year on the N face and 26% of the year on the S face, which can be explained by the close proximity of the river Hermitage Water, 15m away, contributing to a moist microclimate in the valley containing the castle. *T.aurea* experienced varying degrees of water stress throughout the time period and it is suggested that various adaptations of this species such as thick cell walls, production of mucilage and the alga's ability to osmoregulate, could aid survival under variable environmental conditions.

1. Introduction

Green algae have been largely ignored as a contributor to stone decay by most researchers world wide, the tendency being to concentrate on the effects of lichen, fungi and bacteria on stone deterioration. A project was undertaken to investigate the mechanisms of stone decay, in this case sandstone decay, by an orange coloured alga situated on a 13th Century castle in the Scottish Borders, U.K. [1, 2]. The stone was colonised by a biological community predominantly the species *Trentepohlia aurea*. Early work showed *T.aurea* to be growing in a particular pattern on the north facade of the castle and associated with spalling of the stone surface [3]. To assess the influence of environment in terms of humidity and temperature on the distribution of algal growth, climate data monitoring was carried out at Hermitage Castle. This project assessed some facets of the microclimate in which this particular alga was proliferating, and as a result causing decay of the sandstone.

Species of *Trentepohlia* have been observed on a variety of substrates (granite, gneiss, sandstone, limestone, mortar, brick, concrete, painted concrete, wood and smooth ceramic tiles) and under a range of climatic temperature regimes in many different countries both north and south of the equator [1]. The alga thus has an ability to adapt well to many different substrates and humidities. Climate and substrate are recognized as an influence on the distribution of biological growths, for example algae usually becomes well established on damp northern facades, while certain lichens tend to be found more often on south and east facing slopes in drier zones of the substrate. Microclimate studies, which link the environmental conditions around a building with specific types of decay such as freeze thaw and salt crystallisation/wet dry cycling, are more common than studies where the microclimate of a building and its association with biodeterioration is under closer scrutiny [4-7].

2. Microclimate monitoring at Hermitage Castle; background and methodology

Hermitage Castle, currently managed by Historic Scotland, is situated 16 miles away from Hawick in open moorland with low intensity farmland, mainly grazed by sheep. Much of the surrounding vegetation is grass and sedge marshland overlying peat bog. There are no known local sources of pollution and the acidity of the rainfall at Hermitage was measured at pH 4.8, which is the national average for Scotland. The Castle is surrounded by un-investigated earthworks and the south facade slopes down towards Hermitage Water, approximately 15m away, where well established trees grow on the banks. The river is approximately 4m wide and 0.5m deep near the Castle. There is a raised bank approximately 10m high on the west face, and on the north face the earthworks slope steeply to the base of the wall close to the north west tower. Hermitage Castle is made up of four towers and connecting walls situated around a 13th Century manor house. Two facades were chosen for the microclimate study, the North West Tower North face (NWTN face) and the South West Tower South face (SWTS face) as shown in Figure 1. The SWTS face was chosen for the study because the facade had only a small area with *T.aurea* present, for example a sheltered crevice in a single stone. In terms of aspect the SWTS face was the complete opposite to the NWTN face which was heavily colonised with *T.aurea,* and associated stone decay. The geometry of the ruined building on the two faces was such that the data loggers/sensors could be placed high enough up to be out of the reach of the general public, in window recesses, where they were only accessible to restricted personnel.

Envirologgers V632 and a Squirrel 1250 series data logger were used to record data from sensors positioned on the castle walls in order to gain an appreciation of some aspects of the microclimate of parts of Hermitage Castle. The loggers were kept in re-sealable plastic containers with silica gel, to avoid corrosion of the computer interface links. Sensors were placed on the NWTN face (Figure 1) in areas where *T.aurea* was prevalent. Sensors recording the stone surface temperature, globe temperature (combined air temperature and radiant heat), relative humidity and light intensity were placed on both the NWTN face and the SWTS face of Hermitage castle. Holes were drilled into the mortar and wooden plugs and metal supports held the temperature sensors flat against the stone. The globe sensor was positioned approximately 40mm above the stone surface. The relative humidity probes were positioned in window alcoves with the sensor tips protected from direct rainfall by plastic canisters. With regard to stone surface temperature, one sensor was placed on each of 4 stones on the NWTN face and the SWTS face. The stones selected for monitoring on each face are described in simple terms (colour and biological growths) as shown in Table 1.

Table 1. Simple description of sandstones in contact with sensors on Hermitage Castle.

	NWTN FACE				SWTS FACE			
Stone	A	B	C	D	E	F	G	H
colour	white	white	white	white	yellow	yellow	white	white
T Present at sensor tip	yes	yes	no	yes	no	no	no	no
Organic growth	*T.aurea*	*T.aurea*	white lichen	*T.aurea*	grey growth	black growth	grey	grey
Decay ?	spall	spall	no spall	spall	none	none	none	none

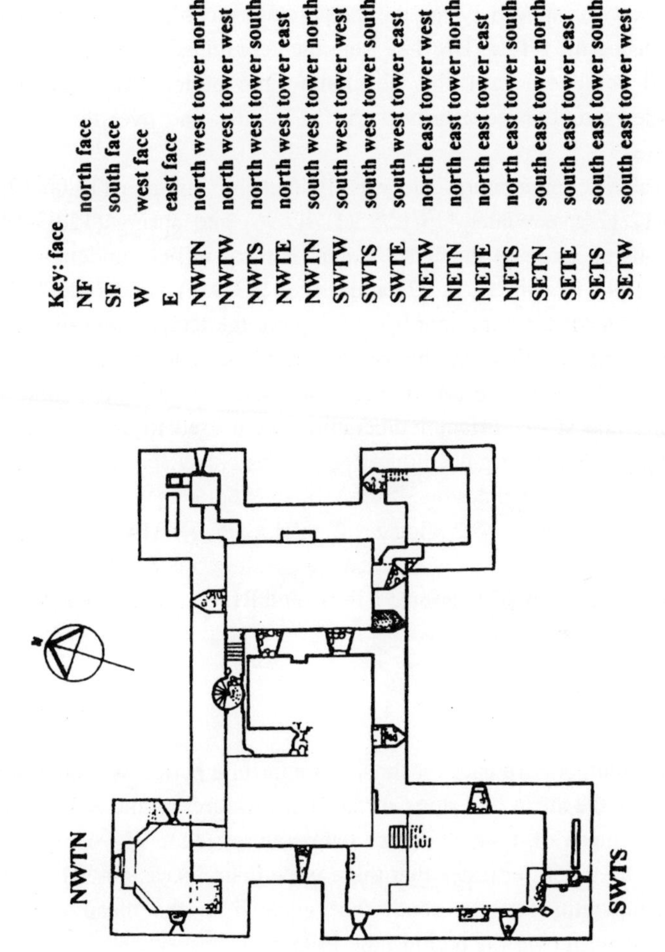

Figure 1. First Floor Plan of Hermitage Castle indicating facades used for data collection (NWTN&SWTS).

Microclimate data was received over a 15 month period from 12/06/95 to 13/09/96. The data presented here was obtained over a calendar year from 12/06/95 to 13/06/96 from sensors recording variables at time intervals of 00:00, 06:00, 12:00 and 18:00 hours each day. Light intensity data could not be included here because of intermittent failure of the light sensor on one face throughout the time period.

Any significant variation between the temperature recorded on individual stones was determined using the t-Test. The data were then combined from the 4 sensors, the mean taken and compared over the 12 month time period. The maximum and minimum recorded temperature data over the 12 months and over the four seasons was also calculated.

The stone surface temperature data was divided into summer (13/06-12/09/95), autumn (13/09-12/12/95), winter (13/12/95-12/03/96) and spring (13/03-12/06/96) which corresponds as close to the U.K. Meteorological Office guidelines (summer {06-08} autumn {09-11} winter {12-02} and spring {03-05}) as possible. Calculation of dewpoint was considered important because where the temperature of the stone is below dewpoint, condensation on the stone can be said to have occurred. This moisture could then be available for the organisms living in or at the stone surface. Relative humidity and stone surface temperatures were used to calculate dew point using the following calculation:

$$\log p_s = 30.59051 - 8.2 \times \log T + 2.4804 \times 10^{-3}\, T - 3142.31/T$$

where $\log p_s =$ vapour pressure. T = °K and RH = $p/ps \times 100\%$ [8]

3. Results and discussion

The stone surface temperature data for the 12 month time period was analysed and a scatter plot showing the mean of 4 stone surface temperatures on the NWTN face, was plotted against the mean of 4 stone surface temperatures (June 95-96) on the SWTS face (Figure 2). The graph indicates that there were many occasions when the mean stone surface temperature on the south face equated to the mean stone surface temperature on the north over the period June 1995-6.

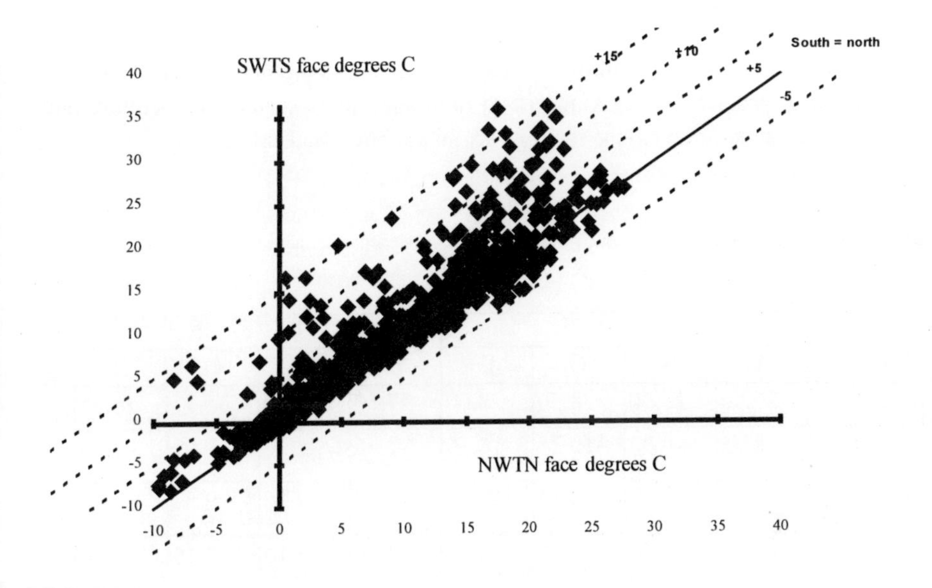

Figure 2. A scatter plot of the mean of 4 stone surface temperatures on the NWTN face and the mean of 4 stone surface temperatures on the SWTS face between June 1995-1996.

Using the t-Test, results on the stone surface temperature data indicated there was no significant difference between the 4 individual sensors recording temperatures on the NWTN face throughout the year and likewise for the S face. There was also no significant difference between the surface temperature of stones where *T.aurea* was present and stone where *T.aurea* was absent on the NWTN face. Stone sizes ranged from $50mm^2$ to $800mm^2$, however, size of stone did not appear to change the stone surface temperature as there was also no significant difference between the larger white sandstone and the smaller white sandstone (Stones G & H on Table 1) on the S face throughout the year. The colour variance between yellow and white sandstone on the S face also did not appear to affect the stone surface temperature significantly, (although many more sensors would be necessary to fully determine whether stone size or colour does affect temperature). The data does however indicate that all the sensors were in agreement, regardless of stone colour and size, consequently the temperature data obtained could be confidently used to compare between the N and S facades.

The t-Test indicated a significant difference ($P <= 0.01$) between the NWTN and SWTS face surface stone temperatures throughout the year. For 72% of the year the SWTS face surface temperature was greater than the NWTN face with the greatest difference being 18.40°C. For 24% of the time the NWTN face temperature was higher than the SWTS face with the greatest difference being 4.40°C. The data was manipulated to produce the maximum and minimum stone surface temperature and the mean and standard deviations for June 95-96 as shown in Table 2.

Table 2. Maximum, minimum, mean stone surface temperatures in degrees C and standard deviation measured on the NWTN face and SWTS face for the year June 95-96 at Hermitage Castle.

°C	NWTN FACE				SWTS FACE			
Stone	A	B	C	D	E	F	G	H
max.	26.00	28.00	28.50	28.00	37.50	37.00	35.75	36.25
min.	-8.75	-9.00	-10.25	-10.50	-7.75	-8.00	-8.00	-8.00
mean	8.41	8.37	8.35	8.29	9.55	9.26	9.62	9.24
s.d.	6.32	6.50	6.53	6.54	7.19	7.10	7.15	7.04

The maximum stone surface temperature recorded in summer on the N face was 28.50°C, whilst the minimum temperature was -10.50°C. The maximum stone surface temperature on the S face was 37.50°C, whilst the minimum temperature was -8.00°C. These results however do not indicate the effects of this temperature on the alga or whether such temperatures affect its growth cycle. However, it is known that the optimum temperature conditions for growth of *Trentepohlia odorata* under laboratory conditions as defined by Ong, Lim and Wee [9] is 25.00°C. From the maximum/minimum temperatures recorded at the castle the algae colonising the stone is clearly tolerant of extremes either side of its optimum range.

Figure 3 shows the mean maximum and mean minimum stone surface temperatures on the NWTN face and the SWTS face. The NWTN face experienced lower mean maximum temperatures throughout the year compared to the SWTS face. The mean minimum temperatures experienced by both faces varied little, apart from during the winter where the NWTN face experienced a lower mean minimum temperature.

Table 3 shows the mean and standard deviation of each stone surface temperature for each of the four seasons of the year on the NWTN face and SWTS face. The higher mean temperatures occur on the SWTS face, whilst only a small difference is recorded between the 2 facades during the spring period. The standard deviation indicates smaller changes of temperature on the north face during all seasons this is also shown on Table 3.

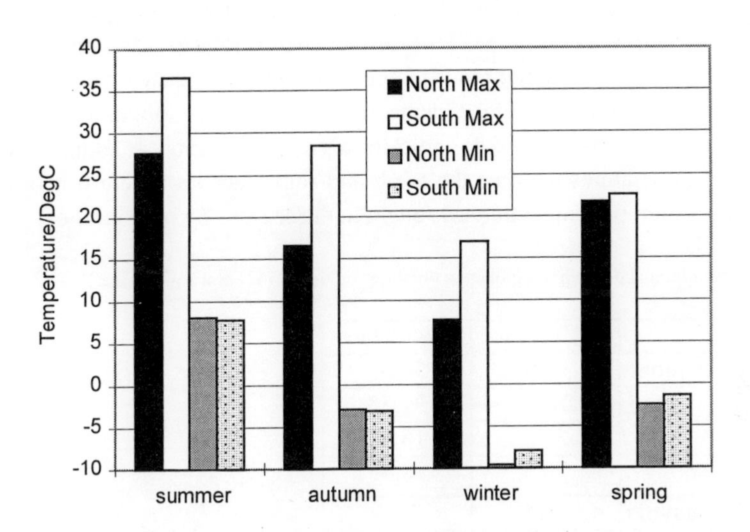

Figure 3. Mean of Maximum and minimum stone surface temperatures on NWTN face and SWTS face for each season.

Table 3. Mean stone surface temperature in degrees C and standard deviation of each by season.

		NWTN FACE				SWTS FACE			
	Stone	A	B	C	D	E	F	G	H
Summer	mean	16.12	16.22	16.02	15.96	17.80	17.64	17.85	17.27
	s.d.	3.51	4.01	4.22	4.22	5.80	5.74	5.64	5.67
Autumn	mean	8.41	8.26	8.32	8.29	9.67	9.52	9.81	9.43
	s.d.	4.00	4.06	4.26	4.29	4.78	4.81	4.91	4.76
Winter	mean	1.79	1.67	1.82	1.74	2.95	2.82	3.04	2.78
	s.d.	7.28	7.29	7.19	7.14	7.74	7.02	7.71	7.42
Spring	mean	4.28	4.46	4.54	4.55	4.69	3.97	4.57	4.58
	s.d.	4.28	4.46	4.54	4.55	4.69	3.98	4.57	4.58

Globe air temperatures were also measured for both the NWTN and SWTS facades. Globe air temperature is a measure of the combined effects of air and radiant temperatures. The mean globe temperature throughout the recorded time period on the NWTN face was 8.2°C (±7.2°C) while the SWTS face mean was 9.3°C (±7.7°C). Although there was no significant difference between the globe and stone surface stone temperatures on the N or S face throughout the recorded time period, the difference between both NWTN and SWTS face globe temperatures was statistically significant ($P<=0.01$), with the higher globe temperature being associated with S face. Table 4 shows the maximum globe temperatures and the minimum globe temperatures for the 4 seasons. The data indicates a greater variation in autumn and winter maximum temperatures between the north and south facades. There was little variation between minimum temperatures throughout the seasons on both facades.

Table 4. Seasonal maximum and minimum globe temperatures on the NWTN and SWTS face.

		NWTN face	SWTS face
Max. °C	summer	36.50	36.85
	autumn	21.25	32.55
	winter	8.50	18.85
	spring	30.5	28.45
Min. °C	summer	5.75	3.15
	autumn	-5.50	-5.30
	winter	-11.50	-12.30
	spring	-4.75	-2.20

Hermitage castle is situated close to the Eskdale Muir weather station, which records the highest rainfall in Scotland (UK Meteorological Office data). *Trentepohlia sp.* survive best in humid conditions, but can tolerate the relatively high salt concentrations of coastal regions (authors observations) and periods of drought. In dry weather, condensation may be the main source of moisture for organisms growing at the stone surface. Some algae and lichens are extremely tolerant of drought conditions and are capable of resuming photosynthesis rapidly after sudden wetting [10]. In addition, low rates of photosynthesis can occur in some organisms which are able to utilise water vapour at relative humidity of 80% [11]. Under such conditions condensation would certainly be an important source of water for these organisms. Where the temperature of the stone is below dewpoint, condensation on the stone can be said to have occurred. The dewpoint was calculated from the data obtained at Hermitage Castle since it was hoped that the results may help explain the distribution of organisms in relation to moisture. The results showed that for 4% of the year, condensation occurred on both the NWTN face and SWTS face at the same time.

Condensation occurred for 11% of the year on the NWTN face and 26% of the year on the SWTS face. The data was analysed to investigate whether condensation occurred over consecutive 6 hour periods and the results are shown below:

Consecutive findings	NWTN	SWTS
1 time (6 hourly)	11%	26%
2 times (12 hourly)	4%	5%
3 times (18 hourly)	3%	1%
4 times (24 hourly)	3%	<1%

Since Hermitage Castle is situated in a valley 15m away from a river, the river could create a moist microclimate around part of the castle. In the northern hemisphere, one would normally expect a higher rate of condensation on the N side of a building than on the S Face of a building (because of insolation), however the close proximity of the river may explain in part, the higher rate of condensation recorded on the SWTS face.

Whilst there is still no occurrence of *T.aurea* on the SWTS face, the alga and spalling were found on the E face of the South West Tower at the beginning of the study in June 1994 (there was no monitoring equipment on this facade). However, on the last visit to Hermitage in June 1996 *T.aurea* had begun to colonise the S face of the castle. Algae generally grows less prolifically on S faces in the northern hemisphere. However, *Trentepohlia sp* appears to be able to survive extremes of temperature, and if moisture is one of the limiting factors on occurrence then the ameliorating effect of Hermitage water could encourage its colonisation on the SWTE face, and explain its recent colonisation of the S Face.

More specifically, on the SWTS Face there was more condensation, there were also more times (32% of the year) when the relative humidity was <90%, a point at which biological organisms can become stressed and stop growing [12]. The lowest relative humidity was 22.5% on the N face of the NWTN face and 30.3% on the SWTS face, again indicating the ameliorating effect of the river (since one would have expected relative humidity to be lowest on the S face). With regard to the NWTN face, 21% of the year, relative humidity was <90%. Algae require moisture to survive and grow and although microbial growth can occur as low as 61% relative humidity in the laboratory [12] most organisms generally require 90% relative humidity or above in which to proliferate. The ability to grow at low water potentials depends on the success of the organism to adapt. Most epilithic organisms, those that grow on the surface of the stone, should be well adapted to survive periods of drought. The production of mucilage, modification of cell shape by adjustment of cell surface to volume ratio and the accumulation of salts or organic solutes in the cell are all strategies which have been adapted by micro-organisms to survive during water stress [13-15, 2]. Some algal species such as *Trentepohlia* may accumulate intracelluar organic solutes such as mannitol to aid adaptability to varying moisture levels [15]. This factor together with the thick striated cell walls [3] are signs that *T.aurea* is indeed well adapted to varying moisture/drought conditions from Hermitage Castle.

The microclimate data accumulated from Hermitage Castle indicates that *T.aurea* is under severe water stress at various periods throughout the year and on a diurnal basis. However, *Trentepohlia sp.* has been reported, growing on sandstone in places receiving only atmospheric moisture (Round 1965, in [16]) however, in Singapore the growth of this algal species was found to be discouraged in the presence of runoff water zones on concrete walls [17], probably due to overgrowth by faster growing species such as cyanobacteria.

The decay mechanism of stone at Hermitage Castle has been largely attributed to physical processes of directed growth by *T.aurea* although some evidence of biochemical activity is present in the decayed stone, the mechanisms of which rely on the presence of moisture [1]. Other workers [18-21] have identified the importance of microclimate (both internal and external) to incidence of biodeterioration, though the work is often linked to studies of air pollution as well. There are few studies like the work undertaken at Hermitage Castle, which have considered the effects of microclimate on biodeterioration which is in a rural setting.

4. Conclusions

The microclimate data have provided some supportive information about the conditions which *T.aurea* experiences at Hermitage Castle, in particular the temperature extremes that the organism is living under. The factors affecting the distribution of alga on the stone at Hermitage Castle appear to be related to microclimate, however, the presence of ditches, sloping ground, Hermitage Water and trees and vegetation are also likely contributory factors. So is the proximity of livestock around the castle which could provide microflora with important organic and inorganic compounds for biological growth. Of course, a further study involving collection of data over a longer time period and a full weather station would be advantageous to assess more fully the microclimate around Hermitage Castle with respect to the growth of *Trentepohlia.*

5. References

1. Jones M S. Wakefield R D. Forsyth G., Biodeterioration of Sandstone by Algae and Application of Stone Conservation Method. Engineering and Physical Sciences Research Council (EPSRC*) Final Report GR:J91500.* (October, 1996).
2. Wakefield R D. Jones M S. Young M E. Nicholson K. Urquhart D C M. Wilson M J., Investigations of Decayed Sandstone Colonised by a Species of Trentepohlia. *Aerobiologia* **12**, (May, 1996), 19-25.
3. Wakefield R.D. Jones M.S. Forsyth G. Decay of Sandstone Colonised by an Epilithic Microbial Community. In *Processes of Urban Stone Decay*. ed. Smith B. J. and Warke P. A., Donhead, (March, 1996), Chapter 8, 90-99, ISBN 1 873394 20 9.
4. Halsey D P. Dews S J. Mitchell D J. and Harris F C., Influence of Aspect upon Sandstone Weathering: The Role of Climatic Cycles in Flaking and Scaling. In *Proceedings of the 8th International Congress on the Deterioration and Conservation of Stone,* ed. Riederer J., (Berlin, Germany, September 30 - October 41996) 849-860.
5. Halsey D. The Weathering of Sandstone, with Particular Reference to Buildings in the West Midlands, UK. Unpublished PhD Thesis, University of Wolverhampton, 1996.
6. Hockman A. Kessler D., Thermal and Moisture Expansion Studies of some Domestic Granites. *U. S. Department of Commerce National Bureau of Standards, RP 2087*, **44**, (1950), 395-410.
7. McGreevy J., Some Perspectives on Frost Shattering. *Progress in Physical Geography.* **5**, 1, (1981), 56-75.
8. Chadderton D V., *Air Conditioning: A Practical Aproach*, (Publisher E & F Spon, 1991), 42.
9. Ong B. Lim M. Wee Y.C., Effects Of Dessication and Illumination on Photosynthesis and Pigmentation of an Edaphic Population of *Trentepohlia Odoratoa* (Chlorophyta). *J.Phycol.* **28**, (1992), 768-772.
10. Potts M. Friedmann E.I., Effects on Water Stress on Crytoendolithic Cyanobacteria from Hot Desert Rocks. *Archives of Microbiology*, **130** (Springer - Verlag, 1981), 267-271.
11. Lang A.R.G. Osmotic Coefficients and Water Potentials of Sodium Chloride Solutions From 0 to 40 degrees C. *Australian Journal Chemistry* **20**, (1967) 2017-2023.

12. Horowitz N H., Biological Water Requirements. In Strategies of Microbial Life in Extreme Environments, ed. Shilo M., (Berlin, Dahlem Konferenzen, 1979), 15-27.
13. Warscheid T. Oelting M. Krumbein W. E., Physico-chemical aspects of Biodeterioration Processes on Rocks with Special Regard to Organic Pollutants. *International Biodeterioration*, **28**, (1991), 37-48.
14. Sterflinger K. Krumbein W E., Multiple Stress Factors Affecting Growth of Rock-Inhabiting Black Fungi. *Botanica Acta*, **108**,(1995) 490-496.
15. Reed R.H. and Wright P.J., Release of Mannitol from *Pilayella littoralis* (Phaelphyta: Ectocarpales) in Response to Hypo-osmotic Stress. *Marine Ecology* - Progress Series, **29**, (1986), 205-208.
16. Hueck-van der Plas E.H., The Microbiological Deterioration of Porous Building Materials. *International Biodeterioration Bulletin*, **4,** (1968), 11-28.
17. Wee Y.C. Lee K.B., Proliferation of Algae on Surfaces of Buildings in Singapore. *International Biodeterioration*, **16**, (1980), 113-117.
18. Caneva G. Gori E. Montefinale T., Biodeterioration of Monuments in Relation to Climatic Changes in Rome between 19th-20th Centuries. *The Science of the Total Environment*, **167**, (1995), 205-214.
19. Gomez-Alarcon G. Munoz M. Arino X. Ortega-Calvo J., Microbial Communities in Weathered Sandstones: The case of Carrascoa del Campo Church Spain. *The Science of the Total Environment*, **167**, (1995), 249-254.
20. Valentini M. Sala G. Torvati A., Investigation on Internal Microclimate of a Confined Historical Building. II International Symposium on the Conservation of Monuments in the Mediterranean Basin. ed. Ott and Zezza (Venice, June, 1994), 263-268.
21. Ortega-Calvo J. Arino X. Hernandez-Marine M. Siaz-Jimenez C., Factors Affecting the Weathering and Colonisation of Monuments by Phototrophic Microorganisms. *The Science of the Total Environment*, **167,** (1995) 329-341.

6. Acknowledgements

We would like to thank the following people and organisations; EPSRC for funding this work GR:J91500. Historic Scotland, P Marsden, The Envirologger Company, P Gedge, I Findlay, S Hawley, M Young and D Halsey. Figure 1 is adapted from original drawings by Historic Scotland.

Stone Weathering and Atmospheric Pollution Network '97: Aspects of Stone Weathering, Decay and Conservation.
Edited by M.S. Jones & R.D. Wakefield © 1998 Imperial College Press.

INITIAL INVESTIGATIONS INTO THE DETERIORATION OF MAYA LIMESTONE MONUMENTS AT UXMAL, MEXICO

L. MALDONADO LOPEZ
Centro de Investigación y de Estudios Avanzados del I.P.N
Ap. 73, Cordemex, C.P. 97000, Merida Yucatán, Mexico
Fax. +52 9981 2917
E mail maldonad@kin.cieamer.conacyt.mx

P. TORRES LORIA
Instituto Nacional de Antropología e Historia,Laboratorio de Biología
Ex-Convento de Churubusco, Mexico D. F.
Fax. +52 5688 4519

D. P. HALSEY
Built Environment Research Unit
School of Engineering and the Built Environment
University of Wolverhampton, Wulfruna Street, Wolverhampton
WV1 1SB, United Kingdom
Fax. +44 1902 322680 E mail. d.p.halsey@wlv.ac.uk

Mayan monuments, such as those at Uxmal, have become increasingly important tourist attractions since the 1930's. The tropical climate, tourist activities and increased industrialisation in the region have caused concern over the preservation of these monuments. Preliminary investigations at Uxmal identified three different weathering crusts formed on the limestone, with certain microclimatic conditions being typical for each crust type. All three crusts were related to the colonisation of the stone by fungi, algae or lichen. Although atmospheric pollutants were present at Uxmal this initial investigation found little evidence of related deterioration, however further research is proposed to examine this more thoroughly.

1. Introduction

Deterioration of Maya monuments in Mesoamerica has caused widespread concern, especially to the Mexican government. Through the National Institute of Anthropology and History (INAH) and the Centre for Research and Advanced Studies (CINVESTAV) the Mexican government is studying the deterioration problem and possible methods of preserving these monuments. Deterioration of any monument is a function of complex interactions between current environmental conditions, weathering history and inherent properties of the stone, making it difficult to understand deterioration problems without studying the site concerned. This paper reports some initial observations on the extent of deterioration at Uxmal (Figure 1).

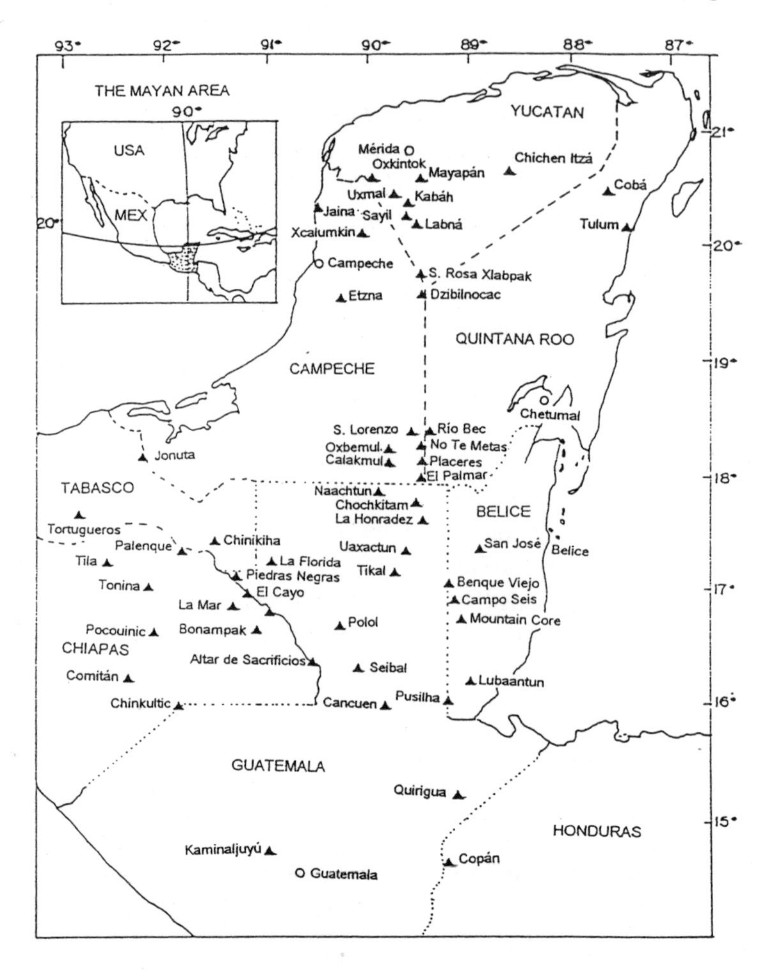

Figure 1. Mayan sites on the Yucatan Peninsula.

The Maya Indians constructed cities with numerous monuments in the Southeast of Mexico and Central America between 150 and 900 AD. As a result, more than 90 restored archaeological sites exist today. These sites are composed of sculptures, monuments and hieroglyphs constructed from local limestones. Some of the best restored are the famous tourist attractions of Uxmal, Chichén-Itzá, Mayapan, Kabah, Sayil and Labná, on the northern Yucatan Peninsula (Figure 1). The cultural importance of Uxmal was demonstrated in 1996 by its inclusion on the UNESCO World Heritage List. Before restoration these sites were half-buried, partly protecting them from deterioration. However, since restoration inscriptions and hieroglyphs have rapidly deteriorated.

In tropical regions archaeological sites typically suffer deterioration due to plant roots, animal activity, high relative humidity, high temperatures and colonisation by micro-organisms. In more recent times tourism and human settlements around archaeological sites have caused significant problems. In 1995 more than 2.5 million people visited Maya ruins located in Yucatan, Campeche, Tabasco and Chiapas, and about 1.5 million visited those located in the Yucatan state only (Figure 1). An increase in exhaust gases from tour buses and emissions from oil wells in the Gulf of Mexico have caused concern that atmospheric pollution may contribute to the deterioration of the monuments in a similar way to that reported in industrialised countries [1, 2].

2. Investigations at Uxmal

Uxmal (latitude 20° 21' 40" longitude 89° 46' 20") is the greatest and most visited Maya ruin comprising of 30 principal groups of structures aligned along an approximately north-south axis (Figure 2). The central part of the city shows a fairly compact arrangement of buildings, constructed between the early-classic and the post-classic periods of the Maya civilisation and abandoned in about 1080 AD [3]. Strongly characteristic of Uxmal is that the buildings are arranged around courts often completely enclosed, such as the nunnery quadrangle (Figures 3 and 4). The historical importance of the city is demonstrated by the fact that, unlike the vast majority of ancient Maya settlements, it never entirely disappeared from history. The old city and its buildings are mentioned in the native chronicles and in other early Maya and Spanish documents [4]. From 1936 onwards the site of Uxmal was gradually excavated, the original stones re-assembled and structural repairs undertaken. There is no record of stone cleaning, biocides and stone consolidants being used at this site.

As a preliminary investigation, the whole site of Uxmal was examined for different forms of stone deterioration, which were photographed *in situ*. Small samples of stone were collected from the monuments and examined with Scanning Electron Microscopy (SEM).

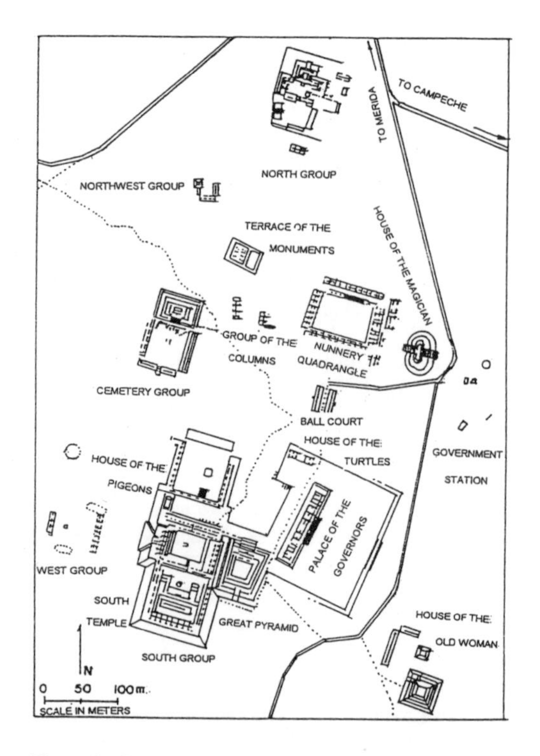

Figure 2. Central section of Uxmal.

Figure 3. Buildings of the Nunnery Quadrangle - northern structure.

3. Observations

The buildings of Uxmal are constructed from white and red-brown limestone blocks about $300mm^2$. In some cases, walls are of one stone type, but in general buildings contain both types. The general appearance of the weathering is very different to that reported for buildings in temperate industrialised countries. The typical black sulphate crusts and areas of white rain washed stone, which dominate limestone monuments in European cities [5], are not present (Figures 3 and 4). However, based on macroscopic morphology and colour, three kinds of crusts were visually delimited and classified as black, white, and orange. Stone type exerted little influence on the distribution of these crusts, but certain microclimatic conditions were typical for each crust type.

Black crusts were present on internal and very sheltered external walls (Figure 5). In these locations the moisture content of the stone is high and constant, and rainfall runoff uncommon. This is similar to the microclimatic conditions described for the formation of black sulphate crusts in European cities [5]. However, a SEM micrograph shows that the surface of the black crust consists of a layer of irregular interwoven filaments (Figure 6). This is distinctively different to black sulphation crusts which typically consist of gypsum and particulate matter [6]. The appearance of black crusts from Uxmal suggests a biological origin and indeed they are similar in appearance to black fungal crusts described for Roman and Greek marbles [7].

White crusts occurred on stones exposed to open-air conditions, which create greater variations in stone temperature and moisture content than the sheltered locations where black crusts were found. These crusts consist of lichens growing as quasi-circular colonies until they overlap and practically cover the surface (Figure 7). Examination with SEM shows the porous array of spheroids that make up the lichen (Figure 8).

In highly exposed locations, such as glyphs-stelas (stone registers of historical events or astronomical data) and roofs, stones exhibit a discontinuous orange crust. The highly exposed position of these stones creates extreme microclimatic conditions, with wetting-drying and heating-cooling cycles being frequent. These cycles will be most frequent in the dry season where wetting occurs during the night, due to condensation, and evaporation occurs during the day as the stone heats up. The macroscopic appearance of the orange crust (Figure 9) and SEM examination (Figure 10) suggest that the crust is an alga from the *Trentepohlia* genus. Similar crusts have been widely reported on other buildings in a range of climates [8].

Figure 4. Buildings of the Nunnery Quadrangle - southern structure.

Fiigure 5. Macroscopic view of black crust.

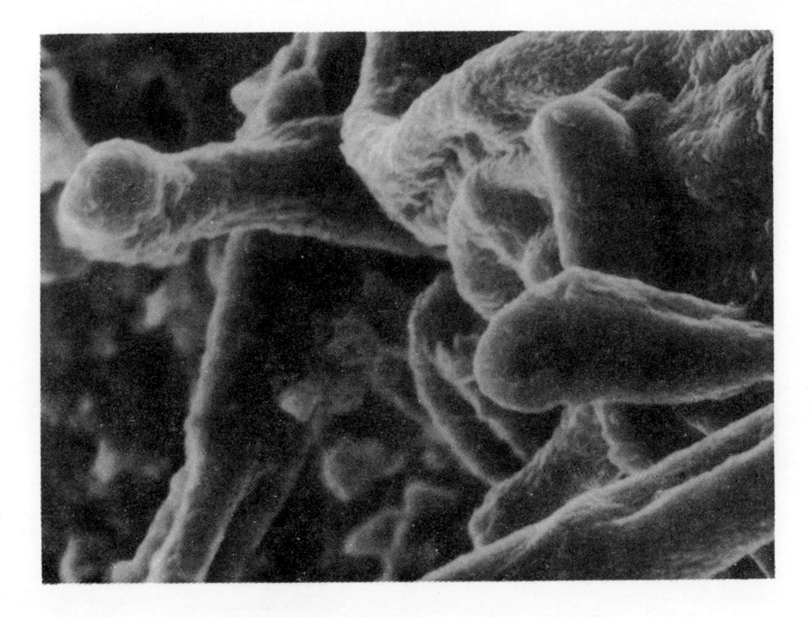

Figure 6. SEM micrograph of black crust (× 1000).

Figure 7. Macroscopic view of white crust.

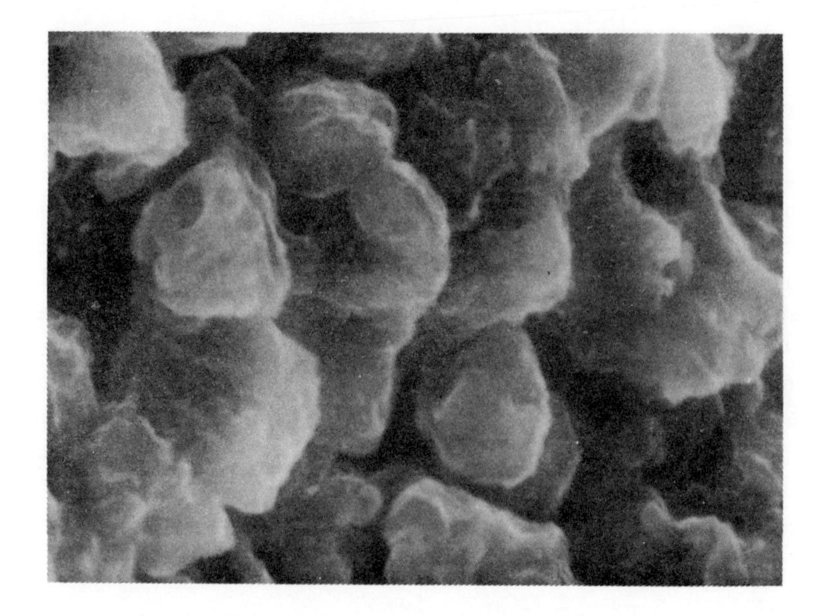

Figure 8. SEM micrograph of white crust (× 3000).

Figure 9. Macroscopic view of orange crust.

Figure 10. SEM micrograph of orange crust (× 1000).

4. Discussion and recommendations

It is well known that lichen, algae, fungi and bacteria secrete a wide range of organic acids, such as oxalic acid, which cause deterioration of stone surfaces [9]. In addition they exert a physical action, due to expansion and contraction associated with the gain and loss of moisture and freeze-thaw cycles [10]. These cycles create physical stresses on the stone which can cause deterioration [11]. Physical mechanisms may be most important for stone affected by the orange and white crusts, as the microclimatic conditions typical for the formation of these crusts may create frequent heating-cooling and wetting-drying cycles. The sheltered location of stone affected by the black crusts is likely to cause very few cycles and reduce the potential of these physical mechanisms to cause stone deterioration. However, the sheltered locations where black crusts are found provide ideal roosts for birds and bats, which create a thick gelatinous coating of excrement on some stones. This provides an additional source for potentially deleterious acids. Inside the buildings this accumulation of organic material, coupled with the ingress of moisture through cracks and joints in the stonework, provides a very humid and potentially deleterious environment. Therefore, excluding birds and bats from the buildings, reducing moisture ingress and removing organic matter may be important steps in conserving these buildings.

The role of sulphur oxides in the deterioration of calcareous stone is well understood [12]. However, further samples of stone are needed to examine sulphate related deterioration at Uxmal. Measurements of pollutants during five months in the Uxmal area, by the sulphation plate technique and the wet candle methods [13], showed that sulphur dioxide and chloride deposition rates were 0.42 mg/m^{-2} day^{-1} and 3.5 mg/m^{-2} day^{-1}, respectively. These values are relatively low compared to European cities. Nevertheless, it is important not to ignore sulphur oxides and other pollutants when investigating the deterioration of Maya monuments, as the potential for pollution related deterioration may be greater in tropical climates than in temperate climates, due to higher humidities.

This preliminary investigation into the deterioration of the monuments of Uxmal shows that biological agents and atmospheric pollution are present at the site. However, further study is required to quantify the extent of deterioration caused by these agents.

5. References

1. Robertson M. G., The effects of acid rain, on Maya ruins. Mesoamerica: *The Journal of Middle America* **1** (1989), 19-30.
2. Halsey D. P. Dews S. J. Mitchell D. J. and Harris F. C., The Black Soiling of Sandstone Buildings in the West Midlands, England: Regional Variations and Decay Mechanisms, in *Processes of Urban Stone Decay*, eds. Smith B. and Warke P. (Donhead, London, 1996), 53-65.
3. Thompson J. E. S., in *Arqueología Maya*, ed. Diana (México, D. F 1977), 79.
4. Saville M. H., Bibliographic notes on Uxmal Yucatán. *Indian Notes and Monographs* **9(2)** (1921), 62-84.
5. Camuffo D. Del Monte M. Sabbioni C. and Vittori O., Wetting, Deterioration and Visual Features of Stone Surfaces in an Urban Area. *Atmospheric Environment* **16(9)** (1982), 2253-2259.
6. Camuffo D. Del Monte M. and Sabbioni C., Origin and Growth Mechanisms of the Sulfated Crusts on Urban Limestone. *Water, Air and Soil Pollution* **19** (1983), 351-359.
7. Gorbushina A. Krumbein W. Hamman C. Panina L., Soukharjevski S. and Wollenzien U., Role of Black Fungi in Colour Change and Biodeterioration of Antique Marbles. *Geomicrobiology Journal* **11** (1993), 205-221.
8. Wakefield R. Jones M. and Forsyth G., Decay of Sandstone Colonised by an Epilithic Algal Community, in *Processes of Urban Stone Decay*, ed. Smith B. and Warke P. (Donhead Publishing, London, 1996), 88-97.
9. Jones D., *Lichens and Pedogenesis. CRC Handbook of Lichenology Vol. 3* (CRC Press Boca Raton, Florida, 1988), 104-109.

10. Fry E. J., Mechanical action of crustaceous lichens on substrate of shale, schist, gneiss, limestone and obsidian. *Annals of Botany* **41** (1927), 437-459.
11. Halsey D. P. Dews S. J. Mitchell D. J. Harris F. C., Influence of Aspect upon Sandstone Weathering: The Role of Climatic Cycles in Flaking and Scaling. in *Proceedings of the 8th International Congress on the Deterioration and Conservation of Stone, Berlin, Germany,* ed. Riederer J. (1996), 849-860.
12. Elfving P. Panas I. Lindqvist O., Model study of first steps in the deterioration of calcareous stone I. Initial surface sulphite formation on calcite. *Applied Surface Science* **74** (1994), 91-98.
13. Standard Practice ISO 9225, *Corrosion of Metals and Alloys-Corrosivity of Atmospheres-Measurements of Pollution.*

6. Acknowledgements

This work was partially supported by the Mexican Consejo Nacional de Ciencia y Technology (CONACYT) under contract 3181A. We gratefully acknowledge the assistance of J. Huchim Herrera (INAH-Yucatan), Guillermina González (Fac. Quim. UNAM), M. Echeverria and G. Casanova.

Stone Weathering and Atmospheric Pollution Network'97: Aspects of Stone Weathering, Decay and Conservation.
Edited by M.S. Jones & R.D. Wakefield © 1998 Imperial College Press.

ALGAL AND NON-ALGAL SOILING RATES OF SOME SCOTTISH BUILDING SANDSTONES

M.E. YOUNG, D. URQUHART
Masonry Conservation Research Group
The Robert Gordon University, Garthdee Road,
Aberdeen AB10 7QB, Scotland, UK.
Tel 01224 263710 Fax 01224 263777
E mail surmy@garthdee1.rgu.ac.uk

Unsoiled, untreated and biocide treated sandstone samples (50x50x15mm) were exposed to natural weathering in an outdoor test rig for four years. Algal and non-algal soiling rates were monitored by measuring lightness and colour changes. On untreated sandstone, lightness changes revealed that most soiling (mainly algal) occurred during autumn and winter coinciding with periods when rainfall, humidity, air pollution and algal growth rates were at a maximum. On biocide treated samples, while the biocide remained active in preventing algal growth, soiling rates were much slower than on untreated samples although again most soiling occurred during the autumn and winter. The rate of soiling (% reduction in lightness) due to the accumulation of algal material could be up to 10% per year, while the rate of non-algal soiling was only about 1% per year. Both algal and non-algal soiling rates were affected by stone type, aspect (north or south) and the original colour of the stone with soiling rates being faster on more porous, north-facing, lighter coloured sandstone. At current soiling rates, projections of soiling rates into the future indicate that untreated surfaces could appear virtually black in 6–22 years. If the biocide treatment remained active on the sandstone then re-soiling to the same level on these samples would take over 30 years. This has important implications for sandstone building facades with respect to stone cleaning methods and the rate of re-soiling after cleaning.

1. Introduction

In addition to being a potential cause of decay in the long term, biological and non-biological soiling of building facades can be aesthetically disfiguring to stone. In order to remove this soiling, many natural stone buildings have been subjected to stone cleaning over the past decades and some buildings have been cleaned a number of times. Re-cleaning programmes are determined, in the main, by the rate of re-soiling of buildings. The nature of the soiling on building stones varies widely. Some studies have found that most soiling is non-biological [1, 2], others have found much soiling to be biological in origin [3]. Although in the past the part played by biological growths in the soiling and decay of building facades was often ignored, it is becoming increasingly acknowledged that micro-organisms are a common and important component of the soiling on many building facades [4]. The composition of the soiling layer has important implications with respect to the rate of re-soiling and for stone cleaning methods.

2. Methodology

To investigate the rates of algal and non-algal soiling on variously treated sandstone samples a test rig (Figure 1) was constructed to hold a number of sandstone samples. The test rig (north/south-facing) was located in an open courtyard at Garthdee, Aberdeen, Scotland (grid ref. NJ 913 030), so that the samples (50x50x15mm) were exposed to natural urban weathering. The sandstone used was chosen to represent a wide range of characteristics typical of building sandstone used in Scotland (Table 1). Samples, which were initially unsoiled, consisted of untreated, grit blasted, chemically cleaned and biocide treated sandstone [13]. This paper is concerned with the soiling rates on untreated (controls) and biocide treated samples as these can be used to monitor algal and non-algal soiling rates on sandstone.

Figure 1. Test rig for sandstone samples

Samples were held at an angle of 60° from the horizontal. The sandstone samples were free of any pre-existing algal growths. To ensure that each stone started with the same potential for algal colonisation each stone sample was initially inoculated with 5ml of a solution containing known algal genera (*Stichococcus, Chlorococcum, Botrydiopsis* and *Tetraspora*) isolated from field samples taken from a range of Scottish sandstones. This ensured that all samples started with the same potential for algal colonisation.

Three biocide treatments were used on the sandstone samples. Two (a quaternary ammonium and an amine based compound) failed within a year. The other (a copper based compound) prevented algal growth for about three years. The results presented here are from samples treated with the copper-based biocide. Failure of the biocide was assumed to have occurred when algal or lichen growth was detected. Samples were not assessed for the growth of any other micro-organisms.

Table 1. Characteristics of Scottish sandstone.

Sandstone	Porosity (%)	L* : initial lightness (%)	Calcareous	Colour	Location (grid ref.)
Blaxter	21.2	67	No	buff	NY 932 902
Cat Castle	14.4	70	No	buff	NZ 010 164
Clashach	18.1	69	Yes	buff	NJ 163 702
Corsehill	17.7	59	Yes	red	NY 207 703
Leoch	6.1	58	Yes	grey	NO 360 359
Locharbriggs	19.6	54	No	red	NX 984 814

Air pollution data on SO_2, NO_x and smoke levels (from the City of Aberdeen Environmental Development Division) were obtained for a single site (Kaimhill, grid ref. NJ 922 037) within 1km of the experimental exposure site, for August 1992 to July 1993. Figure 2 shows data for rainfall recorded over the experimental period.

Algal and non-algal soiling was monitored by measuring changes in the colour of the sandstone samples at intervals using a chroma meter (Minolta Chroma Meter CR-210). The colour scale used, L*a*b*, is a measure of the lightness level (higher value = brighter, lower value = darker), a* and b* are chromaticity co-ordinates, a* for green (-ve) to red (+ve) and b* for blue (-ve) to yellow (+ve). Standard deviation on measurements was ± 0.05. Samples were shielded from incident light and were air dry at the time of measurement.

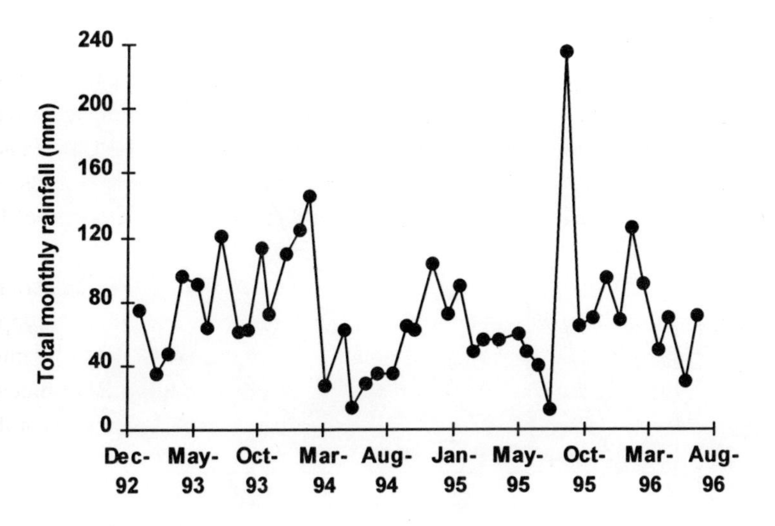

Figure 2. Total monthly rainfall data from Dyce (data from the Meteorological Office).

3. Results

Colour and lightness changes over the period of exposure on untreated and biocide treated Blaxter sandstone samples is shown in Figure 3. Trends on all six sandstones were similar so only those for one stone type are indicated. Changes in algal growth (a*) and lightness levels (L*) on untreated and biocide treated samples showed very clear trends. Colour changes to green showed algal colonisation occurred relatively quickly on untreated control samples with algal growth beginning on biocide treated samples at a later stage. Algal genus' identified were *Chlorococcum*, *Stichococcus*, *Tetraspora* and *Gloeocapsa*. Changes in L* showed reduction in lightness during autumn and winter months coinciding with periods when rainfall, humidity, air pollution and algal growth rates were maximum. Most darkening occurred on untreated samples but darkening to a lesser extent was also observed on biocide treated samples.

On untreated sandstone samples both biological and non-biological soiling could accumulate. On the biocide treated surfaces, while the copper based biocide was active in preventing green algal growths, non-biological soiling along with the un-germinated propagules of micro-organisms and, perhaps, bacteria and fungi could accumulate. No tests were performed to look for the presence of bacteria or fungi on the samples.

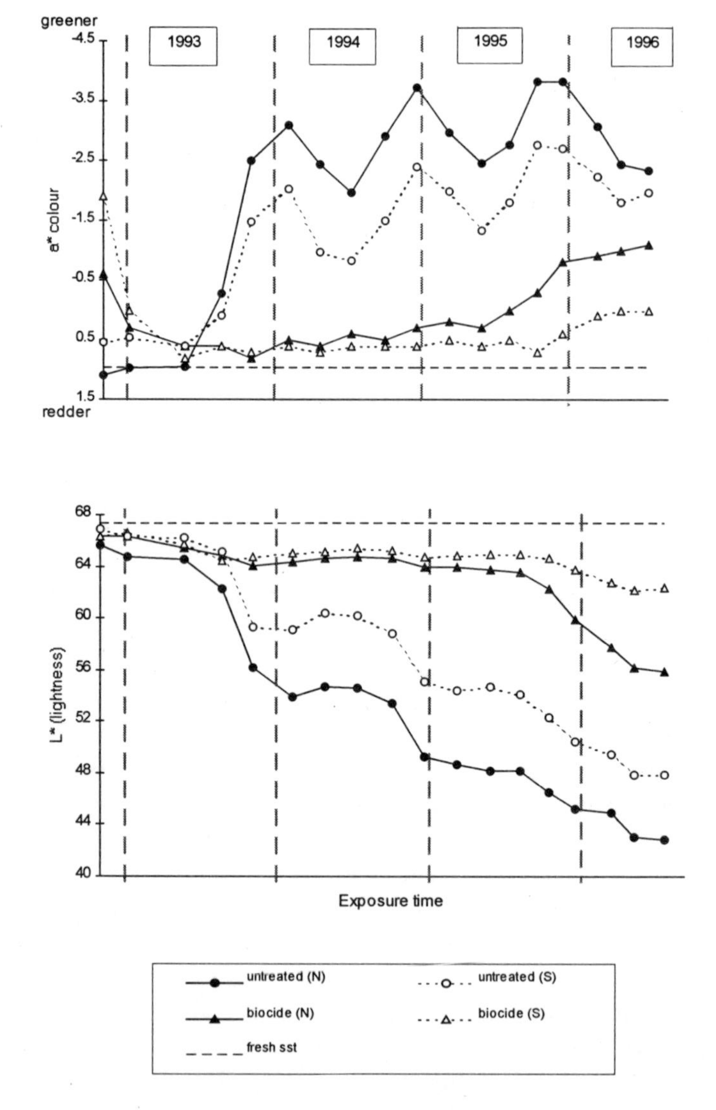

Figure 3 Colour and lightness changes on Blaxter sandstone samples. (N) north, (S) south.

The biocide treatment remained active in preventing algal growth on the sandstone for about 1000 days (Figure 3). Over the first 200 days there were lightness changes associated with loss of an initial blue-green coloration imparted by the copper-based biocide. However, after 200 days, and for as long as the biocide prevented algal growth, any darkening measured on the biocide treated samples was assumed to have been due to accumulation of non-algal and largely non-biological soiling. Comparing measurements of changes in lightness levels (L*) on untreated and biocide treated samples can therefore be useful in the assessment of algal and non-algal/non-biological soiling. Reduction in lightness of surfaces was measured from the spring of one year to the spring of the following year (Tables 2a to c). By 1996 the biocide was no longer preventing algal growth and no further data could be gathered with respect to non-algal soiling.

The data in Table 2a show reductions in lightness (i.e. change in L*) which includes both algal and non-algal soiling. Data in Table 2b show reductions in lightness due solely to non-algal soiling. If it is assumed that the rate of non-algal soiling would be the same on both biocide treated and untreated samples whether or not they also had algal growths present, then it is possible, by subtracting the non-algal component for each sandstone, to obtain a measure of the soiling rate attributable solely to the algal component. This data is shown in Table 2c. It may not be entirely realistic to assume that non-algal soiling rates would be the same whether or not sandstones had algal growths on their surfaces since the growths themselves may affect soiling rates through the production of sticky mucilage or through increased water retention which could affect deposition of particulates or fungal and bacterial cells. However, it is a reasonable assumption that despite these factors non-algal soiling rates should be broadly similar for the two groups.

Regarding pollution levels at the site, data show that SO_2 levels varied from 0-65$\mu g/m^3$ (0-24ppb), annual mean 18$\mu g/m^3$ (7ppb). Within urban areas the United Kingdom Review Group on Acid Rain (UKRGAR) [5] found SO_2 concentrations of about 25-80$\mu g/m^3$ (9-30ppb). NO_x levels at Kaimhill varied from about 6-20$\mu g/m^3$ (3-10ppb), annual mean 13$\mu g/m^3$ (7ppb). The UKRGAR [5] found levels of NO_x across the UK of 8-30$\mu g/m^3$ (4-16ppb). Smoke levels at Kaimhill varied, the highest peaks occurred during winter months with the highest smoke level being 33$\mu g/m^3$. Monthly means varied between 11 and 29$\mu g/m^3$.

Table 2a. Reduction in lightness (L*) measured on untreated sandstone samples. Includes soiling from both biological and non-biological sources.

	Reduction in lightness (%)			
	North-facing		***South-facing***	
Sandstone	**1993-1994**	**1994-1995**	**1993-1994**	**1994-1995**
Blaxter	9.8	6.5	5.8	5.7
Cat Castle	7.9	7.1	3.6	4.7
Clashach	7.0	6.7	3.3	4.3
Corsehill	3.3	3.1	1.1	2.1
Leoch	1.0	1.9	1.5	1.9
Locharbriggs	4.3	3.4	2.4	2.9

Table 2b. Reduction in lightness (L*) measured on biocide treated sandstone samples. Includes only non-algal soiling since the biocide remained active during this period.

	Reduction in lightness (%)			
	North-facing		***South-facing***	
Sandstone	**1993-1994**	**1994-1995**	**1993-1994**	**1994-1995**
Blaxter	0.8	0.8	0.6	0.2
Cat Castle	1.0	0.9	0.7	0.8
Clashach	0.3	0.7	0.2	0.3
Corsehill	0.5	0.7	0.3	0.3
Leoch	1.0	1.0	1.2	0.9
Locharbriggs	0.4	0.9	0.1	0.5

Table 2c. Reduction in lightness (L*) on untreated sandstone samples attributable solely to algal soiling. Obtained by subtracting the values in Table 2b from those of Table 2a.

	Reduction in lightness (%)			
	North-facing		***South-facing***	
Sandstone	**1993-1994**	**1994-1995**	**1993-1994**	**1994-1995**
Blaxter	9.0	5.7	5.2	5.5
Cat Castle	6.9	6.2	2.9	3.9
Clashach	6.7	6.0	3.1	4.0
Corsehill	2.8	2.4	0.8	1.8
Leoch	0.0	0.9	0.3	1.0
Locharbriggs	3.9	2.5	2.3	2.4

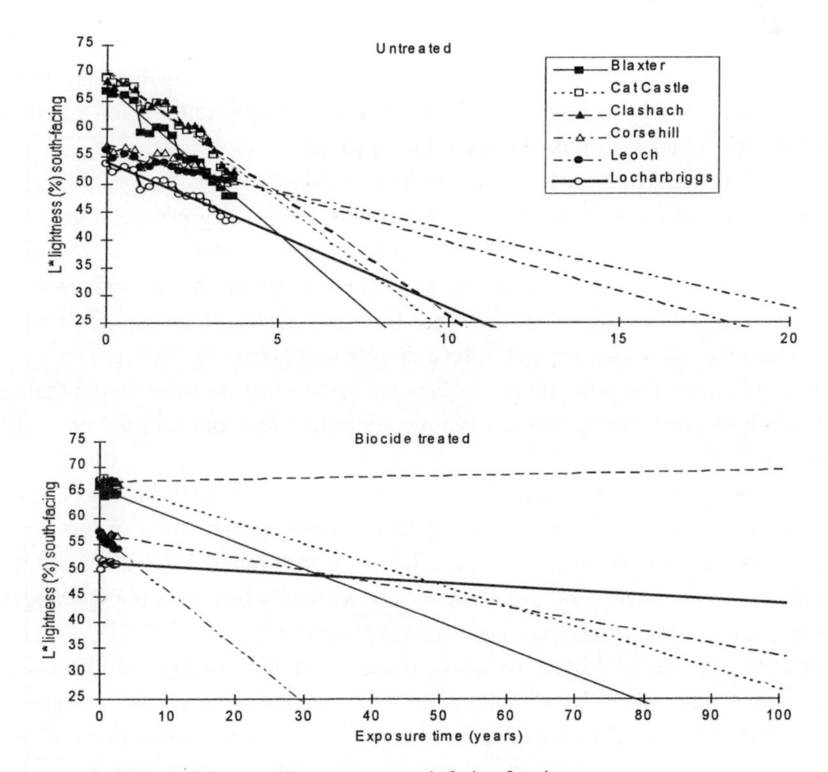

Figure 4. Prediction of future soiling rates on south-facing facades.

4. Discussion

The L* data gave a measure of the rate of what might be called the "visual" soiling of the sandstone surfaces. That is, these data show changes in lightness levels independent of colour changes. Often it is dark soiling rather than green algal growths which is considered to cause soiling of building facades. However, it is not necessarily so that all, or even most, dark soiling on building facades is non-biological in origin. Some organisms are dark in colour [6], some accumulate soiling particles in mucilage [7] and some darken when they die [8]. Darkening is attributable to accumulation of both biological and non-biological soiling.

The data show that the rate of soiling (darkening) due solely to accumulation of biological material derived from green algae could be up to 10% per year while the rate of other dark soiling was only about 1% per year. The rate of soiling on any one sandstone depended on the original colour of the stone (Table 1) and on the amount of algal growth that took place on its surface. Generally, more algal growth occurred on the more porous sandstone (Table 1). The facing direction of the samples strongly influenced the amount of algal growth on their surfaces. Rates of algal soiling were lower on the south-facing samples, which could receive direct sunlight for a period from March until October, than on the north-facing samples.

Leoch sandstone has a very low porosity and was slow to be colonised by algae. The amount of biological soiling on Leoch sandstone, where no green algal growths were observed during this period was, as would be expected, close to zero. The data show that the reduction in lightness due to algal growth in the second year of measurement (1994-1995) was less on north-facing surfaces. However, for south-facing surfaces the rate of soiling in the second year was generally greater. This may be because algal growths were slower to become established on drier, south-facing surfaces, while on north-facing surfaces rapid colonisation took place in the first year of growth.

The differences between rates of dark soiling on north and south-facing samples were not as great as were observed for algal soiling since accumulation of non-algal soiling is not so dependent on the moisture levels in the stones. Particulate soiling does settle rather more readily on damp surfaces so a slightly faster soiling rate might be expected on north-facing samples, and this was observed.

With only approximately four years of data it was not possible to make any definitive predictions about future soiling rates. However, some rough indications may be drawn from the available data by making some broad assumptions. If we assume that the rate of accumulation of soiling will be linear over time then it is possible to extrapolate the available lightness (L*) data into the future. Also, data gathered from research into the cleaning of granite [9], indicated that building stone which appears virtually black in colour had a lightness value of about 25%. Taking this as an end point it is possible to work out how long it would take the untreated and biocide treated samples to achieve this level of soiling (Figure 4). Different sandstone samples soiled at different rates, but if this data can be taken as a broad indication of future trends then untreated north-facing samples would achieve 25% lightness in 6–18 years, south-facing samples in 8–22 years. If the samples could be kept free of algal growths then soiling to 25% lightness on biocide treated samples would take over 30 years. It is probably more likely that soiling rates will not be linear (there is some suggestion of this in the declining soiling rate observed on north-facing samples), and the rate of darkening due to soiling accumulation will decline over time, so that soiling will take longer than is suggested by simple linear extrapolation.

The small size of the samples (50x50x15mm) may mean that the soiling rates measured here are lower than those which might occur on more massive building stones since small samples will dry out more rapidly than larger stones.

5. Conclusions

On average, the algal soiling rate of sandstone samples was found to be eight times faster than the non-algal soiling rate. While this value may hold approximately true for other similar, sloping sandstone surfaces, it should not be assumed to be true of all building surfaces since orientation, porosity, natural stone colour, atmospheric pollution levels and other factors can be important controls on the rates of algal and non-algal soiling.

The results of this study of sloping sandstone surfaces show that under the conditions of this study most soiling is algal in nature and could be prevented by regular re-application of biocides. While on untreated surfaces the sandstones may soil to a virtually black appearance in 6–22 years, on biocide treated surfaces, non-algal soiling could take 30–80 years or more to achieve the same level of soiling (depending on levels of air pollution). These conclusions suggest that regular application of biocides could greatly reduce the need for stone cleaning. However, apart from the obvious difficulties of regular (perhaps every two years) application of biocides, the effects on stone of long term use of biocides are unknown. Some biocides can leave hazardous residues in stone [10], single applications of biocides can cause temporary colour changes to stone [11], some biocides have the potential to affect the clay minerals in stone [12] and the pH of biocides can vary widely and could cause increased dissolution rates of vulnerable minerals.

The rate of re-soiling on sloping sandstone surfaces and other frequently wetted areas can be relatively rapid and on vulnerable stones, chemical cleaning agents containing phosphoric acid can encourage rapid colonisation by algae and lichens [13] which can greatly increase the soiling rate after cleaning.

6. References

1. Nord A.G. and Ericsson T., Chemical analysis of thin black layers on building stone. *Studies in Conservation.* **38** (1993), 25-35.
2. Schiavon N. Chiavari G. Schiavon G. and Fabbri D., Nature and decay effects of urban soiling on granitic building stones. *The Science of the Total Environment,* **167** (1995), 87-101.
3. Ortega-Calvo J.J. Ariño X. Hernandez-Marine M. and Saiz-Jimenez C., Factors affecting the weathering and colonisation of monuments by phototrophic micro-organisms, *The Science of the Total Environment.* **167** (1995), 329-341.
4. Koestler R.J. Warscheid T. and Nieto F., Biodeterioration: Risk factors and their

management, in *Saving our Architectural Heritage: The Conservation of Historic Stone Structures*, eds. N.S. Baer and R. Snethlage. Report of the Dahlem Workshop, Berlin, March 3-8, 1996. (Wiley, Chichester, 1997), 25-36.

5. United Kingdom Review Group on Acid Rain. *Acid Deposition in the United Kingdom 1986-1988*. (Dept. of the Environment and Dept. of Transport) 1990.
6. Saiz-Jimenez C., Microbial melanins in stone monuments. *The Science of the Total Environment,* **167** (1995), 273-286.
7. Lee K.B. and Wee Y.C., Algae growing on walls around Singapore. *Malayan Nature Journal*, **35** (1982), 125-132.
8. Krumbein W.E., Colour changes of building stones and their direct and indirect biological causes, in *Proceedings of the 7th International Congress on Deterioration and Conservation of Stone*, eds. J. Delgado Rodrigues, F. Henriques and F. Telmo Jeremias, Lisbon. Portugal. 1992. (Laboratório Nacional de Engenharia Civil, Lisbon) 1993, 443-452.
9. Urquhart D.C.M. Jones M.S. MacDonald J. Nicholson K.A. and Young M.E., Effects of stone cleaning on granite buildings and monuments. Report to Historic Scotland, Scottish Enterprise and Grampian Enterprise, (1996), ISBN 0 9517989 8 7.
10. Cope B. Garrington N. Matthews A. and Watt D., Biocide residues as a hazard in historic buildings: Pentachlorophenol at Melton Constable Hall. *Journal of Architectural Conservation,* **1**(2), (1995), 36-44.
11. Urquhart D.C.M. Nicholson K.A. Wakefield R. and Young M.E., Biological growths, biocide treatment, soiling and decay of sandstone buildings and monuments. Report to Historic Scotland. (1995), ISBN 0 9517989 9 5.
12. Wakefield R.D. and Jones M.S., Some effects of masonry biocides on intact and decayed stone, in Proceedings of the 8th International Congress on Deterioration and Conservation of Stone. ed. Riederer J., Berlin, Germany. 30th Sept-4th Oct. (1996), 703-716. ISBN 3-00-000779-2.
13. Young M.E., Biological growths and their relationship to the physical and chemical characteristics of sandstones before and after cleaning. Ph.D. Thesis, The Robert Gordon University, Aberdeen (1997).

7. Acknowledgments

The initial stages of this work were undertaken as part of a research project examining algal growth on building sandstone and the effectiveness of biocide treatments. The research programme was funded by Historic Scotland.

Stone Weathering and Atmospheric Pollution Network '97: Aspects of Stone Weathering, Decay and Conservation.
Edited by M.S. Jones & R.D. Wakefield © 1998 Imperial College Press.

MICROSTRUCTURES IN HISTORIC SCOTTISH LIME MORTARS

J.J. HUGHES, P.M. BARTOS, S.J. CUTHBERT,
R.N. STEWART AND J. VALEK
Centre for Advanced Concrete and Masonry,
Dept. of Civil, Structural and Environmental Engineering,
University of Paisley, Paisley, PA1 2BE, Scotland.
E mail hugh-ce0@wpmail.paisley.ac.uk

Detrimental interactions between stone and cement mortars and failures of new lime mortars have focused attention on the microstructures that determine mass transfers which influence interactions within and between masonry and the environment, and thus control rates of decay. The analysis of historic lime mortars is necessary so that compatible replacements can be devised. Calcium carbonate ($CaCO_3$) microstructures in selected historic mortars from 12th-16th century structures in the east and west of Scotland are varied, but can be generalised into four sub-types within a wide continuum; anhedral angular fragments, oval plate-like fragments, dense crystalline masses and amorphous paste structures. The controls on the formation of these structures are not understood, though may relate to variations in carbonation mechanisms and interactions with climate and stone in masonry. There is a clear regional variation in quality for the mortars sampled from the east and west coasts of Scotland in terms of apparent durability and microstructures, suggesting climatic and masonry lithology type interactions as a control on microstructure formation. Post-carbonation carbonate crystallisation is seen in west coast mortars, creating hard durable mortars. The influencing factors on the performance of stone masonry are: age, environment, stone-mortar interactions, binder-aggregate interactions and working and production practices.

1. Introduction

Lime Mortars are a dominant historic masonry binder, having continued in use until the widespread acceptance of cement based mortars earlier this century. Cement hardens more rapidly, has superior durability qualities and requires less on site care than lime. This has led many to believe that in all cases cement is a superior material to lime. Despite these benefits it has become clear that cement possesses physical and chemical characteristics that are incompatible with historic binder materials and with certain stone types. The use of cement has led to physical damage of historic fabric in some cases, and to aesthetic failures in many others.

There is growing evidence for poor performance of stone masonry repaired with cementiteous mortars, but also failures of new "traditional" lime mortars that are currently used for conservation work. This highlights a lack of understanding of the fundamental chemical/mineralogical and microstructural processes that control the architectural and structural performance of lime mortar based stone masonry. In recent decades there has been a worldwide increase in interest in the application of lime mortar, a physically compatible, historically accurate material for conservation and restoration work. Important publications like Historic Scotland Technical Advice Note No. 1 [1] and English Heritage Technical Handbook Vol.3 [2] have begun a rediscovery of the technology of lime use, but such work is largely craft based and the good practice that it advocates involves, as yet, no scientific input.

In an effort to redress this lack of a quantitative background to lime technology, fundamental, interdisciplinary technical research is being pursued which focuses on the influence of microstructures on lime mortar performance and interactions with stone, linked to bulk physical testing on masonry and mortars. The methodology adopted involves analysing microstructures and mineralogy using a range of micro-analytical tools.

The characterisation of historic lime mortars is a key part of this work. It is necessary to measure important parameters such as porosity and permeability in old mortars, especially those which have performed well structurally and architecturally since this defines the physical properties that are useful in a mortar. Historic mortars are also the surviving evidence of historic working practices relating to the production and application of lime mortar. They contain within their structural components and textures, information about potentially beneficial working practices and production methods. However, despite their significance in this respect mortars have not to date been considered a significant conservation objective.

This paper presents preliminary findings from a study of the microstructures of historic Scottish lime mortars. The need to study microstructural parameters is demonstrated through a discussion of the importance of microstructures in controlling the movement of water and gas in masonry. These are the main media for mass transfer in decay processes and also control mortar carbonation.

2. The importance of microstructures for lime mortar performance

Microstructures are important in building materials because they determine the ability of gas and moisture to penetrate a porous substance. Additionally microstructures can affect changes in structure and fluid content and flow which can engender processes of degradation such as salt crystallisation, and control interaction between stone and mortar.

Evidence for the importance of microstructural parameters in the performance of building materials is suggested by the data in Figure 1 which shows cumulative porosity plots derived from Porosimetry, for selected historic mortars and 1 year old cement [3, 4]. Total porosity in cement can be larger than in lime mortars, which is typically around 20%. It is obvious that pore size distribution in cements is very different from lime mortars. The bias to small pore sizes in cements may be the major cause of the damaging interactions seen in practice between cement mortars and stone. Different pore sizes are relevant for different mass transport mechanisms in materials [5]. Small pores exert a greater capillary force and retain water for longer than large pores. Cement mortars draw water to joints because of their small pore size. Evaporation in this wet joint zone is easier from adjacent sandstones with larger pores. Decay in the sandstones is then encouraged due to the stresses encountered during evaporation and the potential deposition of salts. Conversely, lime mortar has porosity that is more compatible with sandstone porosity so water is not localised around mortar joints and evaporation and decay remains evenly spread across the surface of masonry.

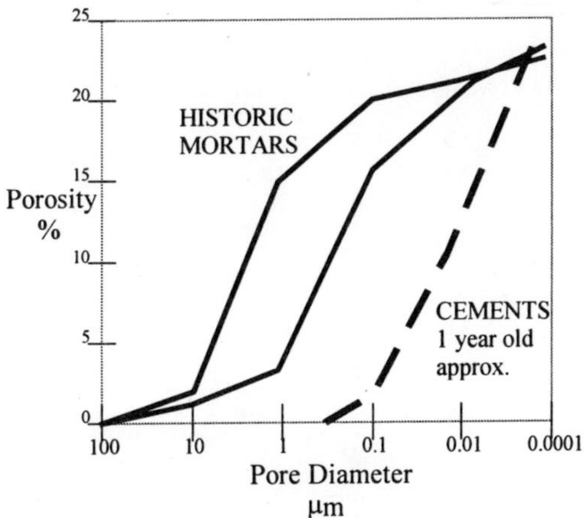

Figure 1. Cumulative pore size comparison between selected typical historic mortars [3], and selected cements [4]. Note greater proportion of larger pores in historic lime mortars.

The detailed nature of the porosity within a lime mortar is especially important because of the process of carbonation. The carbonation reaction is the hardening reaction for lime mortars requiring the ingress of air containing Carbon dioxide (CO_2) into the interior of a mortar mass (Figure 2) and the conversion of lime putty, a stable colloidal suspension of $Ca(OH)_2$, to solid. Carbonation takes place in the presence of CO_2 dissolved in water and depends on a complex interplay of physical conditions such as temperature, humidity, porosity, permeability, CO_2 concentrations [6, 7], and workmanship.

Workmanship not only plays an important part in preventing rapid drying out of fresh masonry, but also in the initial water content and the quality of the manufacturing and mixing processes. Microstructures within lime mortar play a key role in the progress of carbonation as they develop, and are themselves changed by carbonation. The resulting shapes and sizes of lime mortar components control porosity and determine future physical performance and subsequent changes to microstructures due to interactions with stone and the environment.

During lime production $CaCO_3$ undergoes a significant structural transformation. In a limestone $CaCO_3$ can be coarse-grained, massive and crystalline. This is transformed to a porous microcrystalline network of $CaCO_3$ particles in a lime mortar. The process of quicklime production, slaking and mixing of lime putty and lime mortar will have a significant effect on the properties of resultant mortars. The investigation of microstructures in historic mortars could shed light on past processes of mortar production and is a key part of the development of successful new lime mortar mixes.

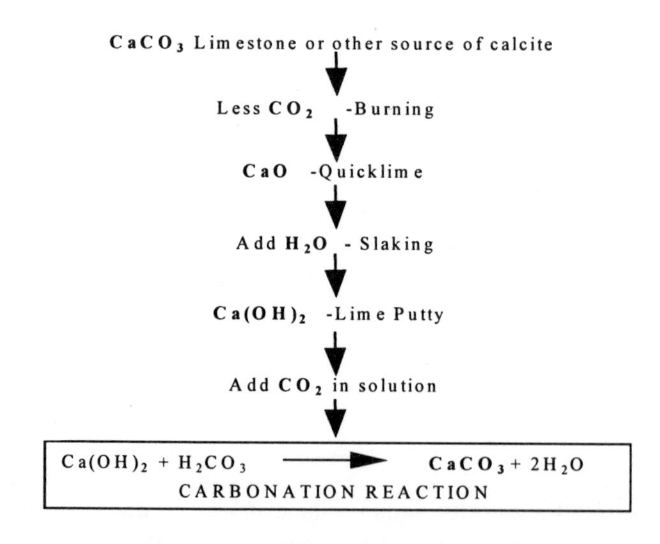

Figure 2. The process of lime production and application. This is commonly known as the 'lime cycle'. The initial form of $CaCO_3$ is usually structurally, if not chemically, distinct from the final carbonated mortar form of $CaCO_3$, hence the linear diagram in this case to emphasise this structural change.

3. Sampling and methods

Samples were taken from eight sites (currently in the care of Historic Scotland), with a wide geographical distribution that are representative of the dominant Scottish lithological masonry types; sandstones (south and east) and granites/metamorphic rock types (north and west). Selected examples (Figures 3-7) were taken from two sites; Inverlochy Castle (west) and St. Andrews Cathedral (east). All material collected had fallen or was removed during restoration of the monuments. Samples, 10mm diameter were examined by Scanning Electron Microscopy (SEM). Secondary electron imaging was used to observe structural elements in the mortars and Energy Dispersive X-ray (EDX) analysis was used to analyse qualitative composition. Polished slices were examined using back scattered electron imaging (BSE) to highlight compositional differences and the nature of porosity.

4. Results and Discussion

Historic Scottish mortars are composed of a binder made of particles of $CaCO_3$, surrounding aggregate grains that can have a wide range of composition relating to local geology or sedimentary environment. Beach and river sands are common, as is shell material. X-Ray Diffraction (XRD) confirm the mineralogy of the binder of the selected samples as calcite, though small quantities (<5%) of other polymorphs of calcium carbonate (vaterite and aragonite) and dolomite ($CaMgCO_3$) could also be present below the detection levels of XRD. Occasionally $Ca(OH)_2$ was present as uncarbonated "lime lumps" like those found by Schouenborg *et al.* [8] along with Si and Mg distributed through the lime binder in unconstrained sites. Cementitious Calcium Silicate Hydrate (CSH) minerals were not observed, so the hydraulicity of the mortars examined is still in question. Gypsum was also present in some samples, but appeared often to be secondary in origin derived from carbonate dissolution and possible interactions with atmospheric SO_2.

Historic lime binders consist of a range of particles usually less than 5μm across with widely varying morphologies. Grain size and shape varied considerably from very small, 0.01μm to large 5μm crystals. Particles were often grouped into clumps or ball like structures linked to other similar structures, that have a fractal, highly self similar nature. Porosity ranged from sub-micron voids to 100μm sized pores and was also fractal. Different pore shapes and size, related to the differences in mortar particles seen and to the number and type of inter-particle contacts. This is also critical in determining the load bearing capacity of the mortar. Particle shape controls porosity and therefore water movement and air permeation into the mortar, and so will control the response to wet/dry and freeze/thaw cycles and carbonation. Four dominant types of mortar particle were common though there was a wide continuum between these types, and some very unusual forms were seen;

- Massive angular fragments. In some cases, possibly remnant unburnt and fractured limestone fragments [9] (Figure 3).
- "Porridge oat" fragments (flat, platy circular/oval shaped fragments). Possibly the carbonated products of hexagonal $Ca(OH)_2$ crystals in colloidal lime putty (Figures 3 & 4). It is not clear if these were formed by the carbonation of wet or dry $Ca(OH)_2$.
- Paste-like material, possibly amorphous with a structure too fine to resolve, often seen at the interface between aggregate and lime, and between lime particles cementing them together. In Figure 4 from St. Andrews Cathedral, it is smoothly cementing "porridge" fragments together and may be related to crystallisation within adsorbed water adhering to aggregate and lime particles.
- Relatively coarse crystalline $CaCO_3$ showing well developed planar crystal faces and cleavage planes (Figure 5). These may relate to crystallisation from saturated solutions of calcium carbonate either due to high initial water contents or later dissolution and reprecipitation. This crystallisation process was suggested by Perander and Raman [10]. Continual cycles of wetting and drying may drive continual dissolution and precipitation of calcium carbonate within a mortar resulting in a dense, crystalline binder.

Isopachous fringes of acicular $CaCO_3$ crystals growing perpendicular to a pore surface can occlude free pore space and close pore throats, indicating the prolonged presence of carbonate saturated pore waters (Figure 6). This has the effect of strengthening the mortar paste, as well as changing the pore volume and permeability of the mortar. In these examples these crystals have formed *after* carbonation of the lime mortar. It is not the result of initial carbonation in a wet environment, but an ageing effect similar to post depositional diagenesis in detrital and carbonate sedimentary rocks. Only old mortars exposed to repeated wet/dry cycles for long periods of time may exhibit such features. The mortars in Figures 3, 5 & 6 are from Inverlochy Castle, built of impermeable igneous and metamorphic rock types, in a high rainfall area. The mortars have been uncovered since the mid 17th century.

Little evidence for a transition zone between lime binder and aggregate has been seen to date. However the structure shown in Figure 5 is located in a relatively dense zone in lime binder within 50-100μm of an aggregate grain and shows little porosity indicating possible evidence for high water contents within the paste at some time.

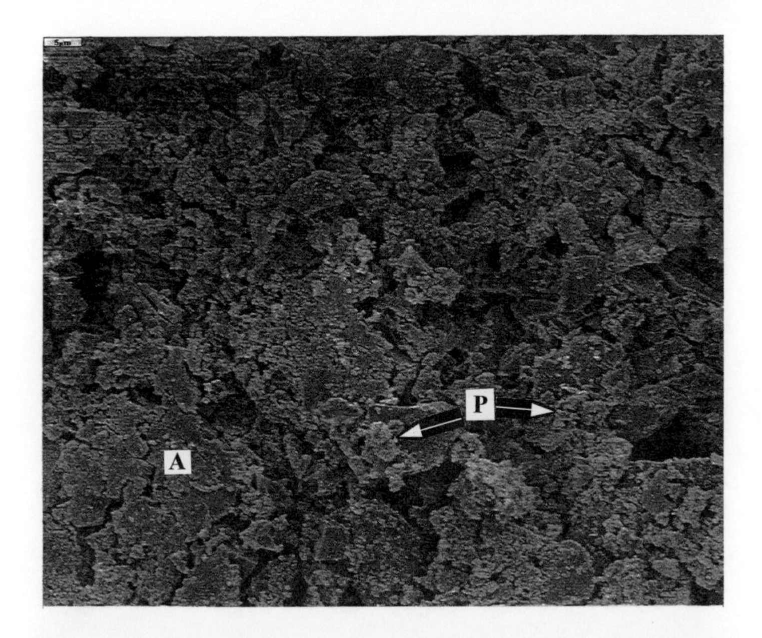

Figure 3. SEM secondary image of a fracture surface from a historic lime mortar from Inverlochy Castle. Angular fragments (A) and "porridge" fragments (P). Field of view is approximately 80μm across.

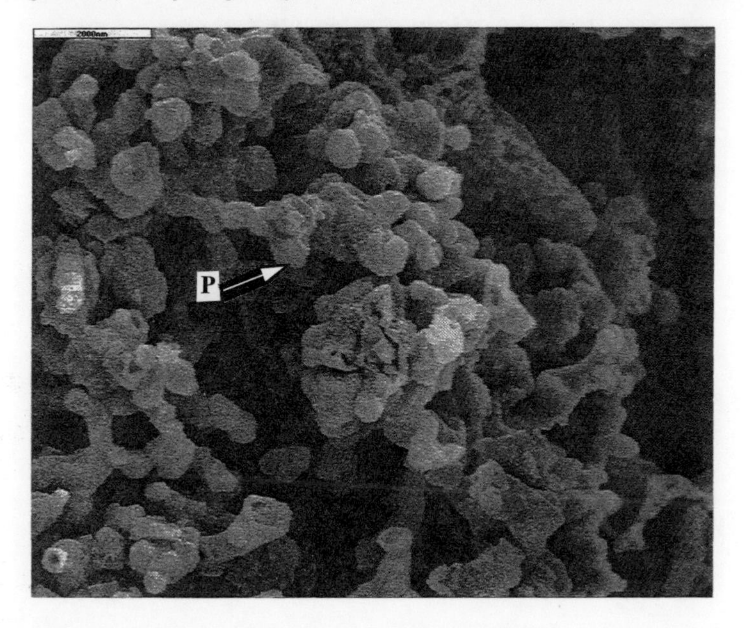

Figure 4. SEM image of a fracture surface from a historic lime mortar from St. Andrews Cathedral. Small "porridge" lime particles (P) cemented together by amorphous lime paste. Field of view approx. 11μm.

Figure 5. SEM image of historic lime mortar from Inverlochy Castle. Rhombohedral calcium carbonate in a dense anhedral microcrystalline binder. Field of view approximately 6μm.

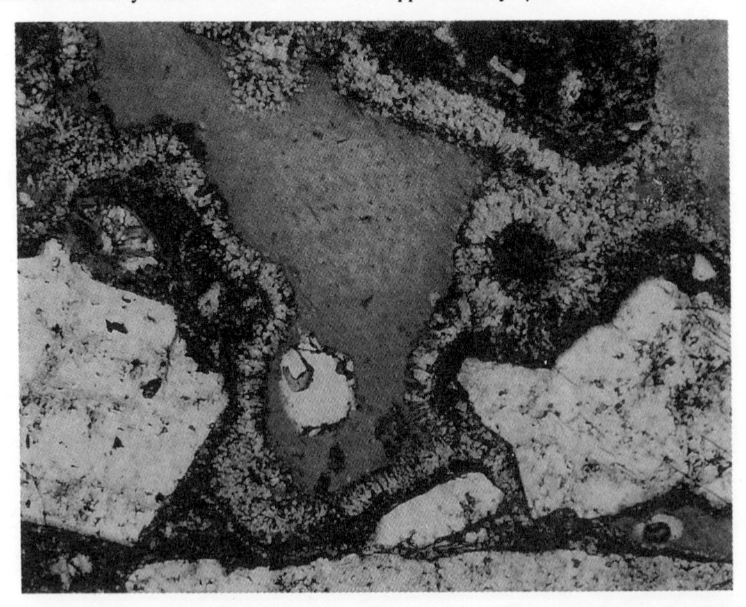

Figure 6. Thin section photograph, plane polarized light, field of view 1μm. Sample from Inverlochy Castle showing isopachous carbonate precipitation in pore space. Note closed pore throat at top right (A).

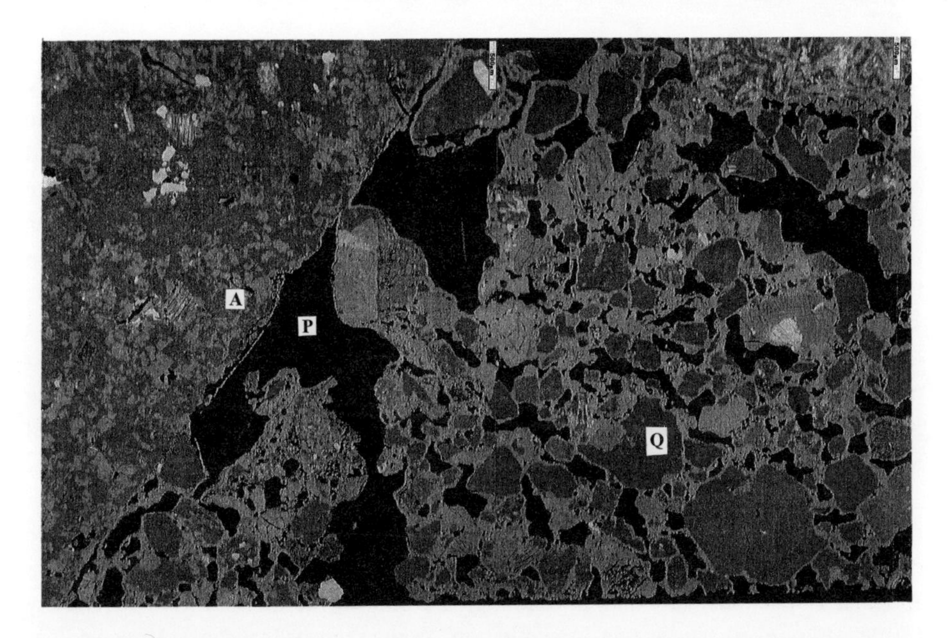

Figure 7. SEM Backscattered Electron image (BSE). Polished section of 12thC. mortar from Inverlochy Castle. View 0.5cm across. Grey contrast reflects composition: light areas have high atomic numbers. Porosity (P), quartz aggregate (Q), lime binder (L). Large igneous aggregate at left (A). High porosity, low binder /aggregate ratio and poor mortar bond with large aggregate. Porosity is not fracture porosity as large pores are enclosed.

Conversely, in another sample an increase in apparent porosity was noted in lime binder on approach to an aggregate interface that may be analogous to transition zones in cement. This type of structure will be an important pathway for fluids and gas in the mortar. It will also localise stress and will be a favoured zone for salt crystallisation and lime dissolution. However the processes that control the formation of such structures in lime mortar are not understood, but may relate to mixing methods and water content.

Many of the historic mortars analysed suggest that high binder/aggregate ratios were common in the past, but the variability is high. Figure 7 shows high porosity in a lime mortar resulting partly from an obvious low binder/aggregate ratio. The large pores may also result from incomplete beating or mixing resulting in air entrainment, however, all grains were well coated in lime binder.

4.1 Regional mortar variation

A qualitative geographical difference in the quality of historic Scottish lime mortars has been observed. Mortars from structures such as Castle Sween and Inverlochy Castle on the west coast of Scotland are very hard, adhere very strongly to masonry and appear to be very durable. East coast mortars, for example from Tantallon Castle and St Andrews and Elgin Cathedrals, have comparatively friable mortars. Both hard and soft mortars are fully carbonated, suggesting some control or interaction with the differing environments for mortars both in terms of climate and of masonry type. Impermeable rock types such as granite and metamorphic rocks may slow carbonation to allow development of a more interlocked and more crystalline structure. Permeable sandstone masonry in the east of Scotland may result in more rapid carbonation and a less well interlocked structure. Regional climatic contrasts may indicate that east coast mortars are more likely to dry out in a less humid and windier climate, resulting in poor carbonation, compared with mortars in the wetter and more humid west coast of Scotland.

The macroscopic difference observed in the mortars sampled for this work can be reflected in their microstructure. The harder west coast mortars often have a more dense structure and contain more angular lime fragments and crystalline $CaCO_3$. Acicular isopachous carbonate pore linings are seen in the mortars from Inverlochy Castle (Figure 6), indicating precipitation from saturated waters. However, as discussed above, this is a late, secondary textural change, not related to initial carbonation of the mortar. Continued wetting of mortar results in carbonate dissolution, movement and subsequent precipitation similar to leaching and stalactite formation seen on many structures. However these historic samples showed no exterior carbonate crusts or skins implying that secondary carbonate crystallisation in these mortars occurred internally.

The east coast mortars are composed dominantly of "porridge oat" fragments and have greater apparent porosity than the west coast samples. Though it must be stressed that these observations are qualitative at the time of writing and will be subject to further testing, especially image analysis, porosity and permeability measurements. There is no hard and fast classification possible as yet, because too few samples have been analysed and there are apparent exceptions for example in Figure 7, the hard west coast mortar has high porosity, though here it relates more to binder/aggregate ratio and possibly poor mixing and not to intra-binder microstructure formation.

The variation found in quality of mortar may be related to the composition of limestone used for lime production in different areas. However, the composition of Scottish limestone is not very variable [11]. Indeed it is Midland Valley limestone which has a higher Si content and is therefore more likely to produce a hydraulic lime which would set more quickly, and have higher strength than the purer metamorphic limestone sourced building limes found north of the Highland Boundary Fault (e.g.

Ballachulish limestone: $CaCO_3$=95-97% [11]). The potential influence of different limestone types is unclear at present and is likely to be complex. The local environment and the nature of lithologies used in masonry are likely to be more critical in controlling the development of microstructures in lime mortars and their subsequent performance. The evident difference between east and west coast mortars may be due mostly to the secondary precipitation of carbonate in pore spaces within the west coast mortars. This is caused by higher rainfall leading to a higher water content, which is localised in mortar joints because the masonry materials are impermeable igneous and metamorphic rock types. Mortars in sandstones buildings, which do not localise moisture in joints to the same extent as the west coast lithologies, do not in the samples so far studied, exhibit microstructures that result in hard durable mortars.

5. Conclusions

There is an urgent need for a better understanding of the fundamental processes involved in the production and application of lime mortars for the conservation of the built heritage, especially in stone masonry structures. Little scientific input has been made to improve methods on site.

Microstructures in historic and modern lime mortars are important because they control the movement of fluid and gas which determine mechanisms of decay, interactions with stone and carbonation in lime mortar. It is carbonation which determines microstructural development and therefore strength and durability. The detailed nature of the microstructures in a lime mortar will be related to strength, in bulk and internally, between binder and aggregate. Microstructures will also relate to bond strengths with stone and brick and will affect the structural performance of masonry. Study of more historic lime mortars is necessary in order to determine desirable qualities in a mortar, and how these relate to various environmental and production process factors. The results will influence and optimise current production procedures and on site practice.

Microstructures in historic Scottish lime mortars are varied, suggesting a number of possible controlling factors operating on their formation. The production process for lime mortar including burning, slaking, storage and mixing may have an important effect on the structures that develop, and therefore the response of lime mortar to environment and its subsequent performance. However these factors are unconstrained at present. Some evidence for variations in burning and mixing can be postulated from historic lime mortars. For example, unburnt limestone fragments in mortar and high porosities relating to low binder/aggregate ratios, and incomplete mixing resulting in large air entrainment pores. Some of these features may have been deliberately engineered by past builders, or could be merely accidental. Nevertheless, it is possible that some of these characteristics may be beneficial.

Evidence for secondary carbonate crystallisation in pores in some mortars highlight the role of environment and masonry type in the development of microstructures through time, especially porosity and permeability. Prolonged aging and exposure to wet conditions has resulted in highly durable and very well cemented mortars in the sampled structures from the west of Scotland. The factors that affect initial carbonation in the short term may have less influence over the long term on the durability and physical properties of historic mortars. This means that we may never be able to accurately reproduce some historic mortars as we find them today, because complex aging processes have operated in some cases over centuries.

Figure 8 summarises the interactions between factors that effect the performance of lime mortar based stone masonry. Continued investigation of microstructures in historic and new conservation mortars, integrated with practical trials of full scale masonry promises to provide much needed scientific evidence to underpin the current renaissance in the application of lime mortars in conservation.

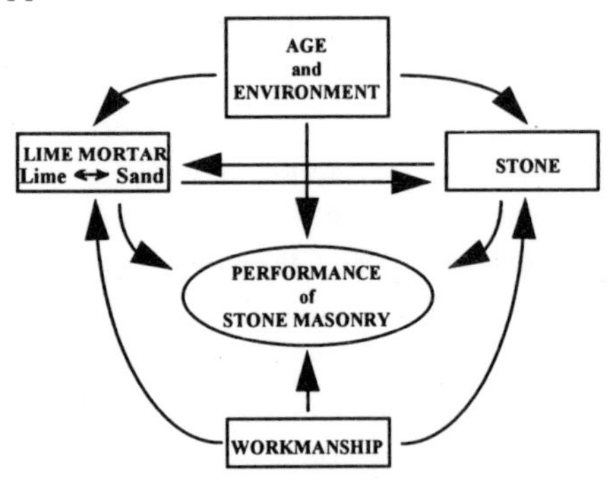

Figure 8. Stone-lime mortar masonry interactions and factors that influence performance.

6. References

1. Gibbons P., Preparation and Use of Lime Mortars. *Technical Advice Note No.1*, Historic Scotland, Technical Conservation, Research and Education Division. (1995).
2. Ashurst J. Ashurst N., Mortars, Plasters and Renders *English Heritage Technical Handbook*, Vol. **3**. (Gower Technical Press, 1989).

3. Thomson M.L. and Basten S.J., Durability of 16th Century Mortar: Arctic Canada versus Inland Ireland, *Proceedings of the Seventh Canadian Masonry Symposium, Hamilton Ontario*, (1995), 678-688.
4. Feldman R.F., in Effects of Fly Ash Incorporation in Cement and Concrete, *Mat. Res.Soc.*, University Park, PA, USA, (1981), 124.
5. Meng B., Determination and interpretation of fractal properties of the sandstone pore system, *Materials and Structures*, **29**, (1996), 195-205.
6. Van Balen K. and Van Gemert D., Modelling lime mortar carbonation *Materials and Structures*, **27**, (1994), 393-398.
7. Moorehead D.R., Cementation by the Carbonation of Hydrated Lime, *Cement and Concrete Research*, **16**, (1986), 700-708.
8. Schouenborg B. Lindqvist J.E. Sandstrom H. Sandstrom M. Sandin K. and Sidmar E., Analysis of Old Lime Plaster and Mortar from Southern Sweden - *A contribution to the Nordic seminar on building limes. Swedish National Testing and Research Institute, Building Technology*, SP Report **34** (1993).
9. Stewart R.N.T. and Bartos P.J.M., *Lime Mortar/Masonry Interaction*, Unpublished Report for the Technical Conservation, Research and Education Division, Historic Scotland, University of Paisley, (1996).
10. Perander T. and Raman T., Ancient and modern mortars in the restoration of historic buildings. *Technical Research Centre of Finland, Research Notes* **450**, (1985).
11. Robertson T. Simpson J.B. Anderson J.G.C., The Limestones of Scotland, Institute of Geological Sciences, *Mem. Geol. Surv.*, Special Reports on the Mineral Resources of Great Britain, **35**, (1949).

7. Acknowledgments

The collaboration of the Technical Conservation, Research and Education Division of Historic Scotland is gratefully acknowledged, both for financial assistance and for continued cooperation in access to historic sites, in obtaining samples and in formulating objectives. The collaboration, advice and support of the Scottish Lime Centre Trust and Masons Mortar Ltd is fundamental in pursuing this research. Margaret Corrigan is thanked for her assistance in the operation of the SEM in the Dept. of Electronic Engineering and Physics at the University of Paisley.

Stone Weathering and Atmospheric Pollution Network '97: Aspects of Stone Weathering, Decay and Conservation.
Edited by M.S. Jones & R.D. Wakefield © 1998 Imperial College Press.

SCOTTISH MARKET CROSSES: TOWARDS THE DEVELOPMENT OF A RISK ASSESSMENT MODEL

L.J. THOMSON, D. URQUHART
School of Construction, Property and Surveying,
The Robert Gordon University, Garthdee Road, Aberdeen, UK, AB10 7QB,
Tel. 01224 263710
E mail surlt@garthdee1.rgu.ac.uk

Historical evidence indicates that the market cross was erected from the twelfth century as the symbol of a burgh's right to trade and was located centrally in a town's market place. Although many examples survive these continue to be under threat from the urban environment and sometimes from inappropriate intervention. This paper presents a review of the past and present significance attached to Scottish market crosses and the risks to which they are exposed. The research described here will assess the condition and analyse the erosion and conservation of this monument type. It is planned that an appropriate recording method will be developed in the course of this, from which the stone surfaces will be mapped digitally. The ultimate aim of the project is to recommend management strategies for the future based upon a risk assessment model. This paper represents a background to, and description of, research currently underway.

1. An introduction to the market cross

1.1 Function

Historically, the market (Scots='*mercat'*) cross was a symbol of a burgh's right to trade and was located centrally in a town's marketplace. In Scotland, documentary evidence indicates that this monument type existed by at least the twelfth century, and it is thought that the early examples were made of wood. The surviving examples are of stone, many dating from the sixteenth and seventeenth centuries, but there are also several more architecturally elaborate examples from the Victorian era and later. Some burghs were recorded in the past as having more than one market cross simultaneously, according to the produce sold around them (e.g. the so-called 'Fish Cross' and 'Flesh Cross' in Aberdeen).

Documentary sources refer to all manner of announcements, celebrations and grisly punishments which took place at market crosses, due largely to the civic associations and the prominent, central location of these monuments. Victims were flogged, branded, burned, hanged or placed in the stocks next to the market cross. Other devices of entrapment, such as the branks and jougs, were often anchored to the market cross itself. This element of their function has prompted an evocative description of their site as "the dreaded theatre of public punishment and shame" [1]. At times, they provided the ideal site for quite historically significant occasions. During royal visits, for example, they were used as ceremonial platforms and as centre-pieces for festive decoration. They were the place from which royal and burghal edicts were proclaimed. The presence of this human factor in the equation colours the history of these monuments: a 'humanistic' value can be added to our appreciation of this monument type. The market cross had a special role within the social context of the town and today it is a symbol of burgh heritage, the location often still marking the historic heart of burgh activity.

1.2 Form

Morphologically, early examples of the market cross were the simplest in construction. These consist generally of a polygonal shaft crowned with a capital and finial, rising from a solid, stepped base, as illustrated by the example at Doune, near Stirling, shown in Figure 1. Depending on the funding available in the burgh for their construction later examples were sometimes more elaborate, such as the round, tower-based market crosses of which five still stand. These consist of a shaft crowned with a capital and finial, as described above, but surmounting an under-structure. This would normally incorporate a doorway and internal staircase, providing access to an elevated platform from which proclamations could be made, as exhibited by examples in Edinburgh and Elgin. The Aberdeen Castlegate example, as shown in Figure 2, originated as such but was converted into an open, vaulted structure, and therefore access to the top is no longer possible. With market crosses from more recent centuries, the fashion often seems to have favoured a square-shaped pedestal base, occasionally a tiered construction, and sometimes with quite elaborate carvings. Figure 3 shows Cullen market cross in Aberdeenshire, of typical Victorian, Gothic Revival design, with pinnacles and crocketting.

Heraldic beasts are common subjects for the finial of market crosses, thus a unicorn or lion can often be seen crowning these monuments. Armorial bearings are another common feature of market crosses, thus heraldic shields frequently adorn the capital or can be seen clasped by the crowning animal. No doubt these devices were intended to visually reinforce the legitimacy of civic authority over the community, perhaps another element in the cultivation of deference proposed in a recent study of the symbolism of late medieval English town halls [2].

Figure 1. Doune Market Cross, Stirling.

Figure 2. Aberdeen Castlegate Market Cross.

Figure 4. Distribution of surviving market crosses in Scotland.

Figure 3. Cullen Market Cross, Aberdeenshire.

1.3 Location

In Scotland today market crosses are generally distributed in town centres, mostly in the eastern and southern areas of the mainland, as shown in Figure 4. This distribution largely reflects the existence of burghs historically. Then, as now, the Highlands of Scotland were less populated than the Lowlands, and indeed the nature of settlement in the Highlands conformed to a different set of economic and social values. There are approximately 145 market crosses surviving in Scotland today, including some twentieth century examples and others which survive as fragments. There are many more recorded by documentary evidence of which the remains no longer survive.

The need for research arises from the lack of recent scholarly focus upon this monument type, the physical risks to which they are exposed and the need for detailed recording of their condition, detailed below.

2. Previous enquiry into market crosses

Towards the end of last century there emerged a surge of interest in Scottish market crosses. This was stimulated by the erection of the present, reconstructed Edinburgh market cross. Discussion about this type of monument took place in certain journals and a call to gather information about destroyed and surviving examples prompted readers to send in details of their local market crosses [3, 4]. This culminated in the publication of a catalogue of Scottish market crosses with an introduction exploring their history and stylistic development in 1900 [1]. In 1914 a Belgian publication highlighted the similarity of Scottish market crosses to Belgian *perrons* [5], and another paper referred to the traditions surrounding boundary stones in an attempt to elucidate the origins of the market cross [6]. In 1928 a booklet discussed the fate of the former Glasgow market cross and explored the possible origins of these monuments in general, and the author pointed to areas of work that he felt needed to be pursued: recording, especially through photography, the exact location of surviving examples, conducting a literature search for individual examples, and exploring the evidence from sources such as local traditions and municipal seals [7]. These lines of enquiry were never followed up. Subsequent publications in the areas of Scottish late medieval burgh life and historical architecture have dealt only in passing with market crosses, outlining their functional and stylistic aspects.

3. Risks to market crosses

The market cross today is frequently still located in town centres and has thus been vulnerable to a range of damage types additional to, and no doubt aggravating, the natural weathering processes. Human agencies such as vandalism, re-siting, well-meaning but inappropriate conservation methods and materials, motor vehicle emissions and other atmospheric pollutants can all cause damage to the stonework. Although there are a few modern examples of Scottish market crosses which are partly or even wholly of granite, they are traditionally carved from locally-quarried sandstone, a material which can be readily sculpted by both artificial and natural agents. Sometimes the inclusion of incompatible materials such as ferrous attachments and cement mortar also aggravates decay. Many market crosses have been destroyed or moved around, sometimes to inappropriate sites, over the years in the course of urban redevelopment. The multitude of local authorities involved in the care of these monuments across Scotland may in the past have led to misguided intervention and mishandling, albeit good-intentioned, by various persons. The crosses are usually *Listed*, and occasionally even *Scheduled* and subject to development control by legislation, but this does not guarantee good management practice for every monument, for example, in techniques of removal or conservation. Historic Scotland is currently embarking upon research into carved stone decay in Scotland and, while some of the observations arising from this may be applicable to market crosses, this particular monument type is not included in the sample as no examples are in State care [8].

4. Recording needs

The development of a method for recording in detail the condition of the stonework of market crosses is required. The survey of historic buildings and monuments has often been more concerned in the past with recording details of architectural form rather than the condition of stone surfaces. Amongst such records there are generally no mapped illustrations to detail the forms and distributions of decay interpreted across the stone surfaces. Detailed records regarding the management and condition of market crosses can seldom be found. National and regional bodies involved in managing buildings and monuments have great workloads and very limited resources. However, the need for more detailed records of monument condition and of conservation activity is now being increasingly recognised. Recently, around fifteen market crosses have undergone State intervention, and been subject to an increased level of photography and more detailed data in the form of conservation reports.

A pioneering classification and mapping technique has recently been developed which records stone condition in great detail using a complicated reference system and which requires quite a high level of geological expertise [9]. While this is a powerful method, one has to consider that many stonework interpreters might be limited in experience and resources, thus it would be useful to develop a system that can be accessible to the maximum amount of people. Indeed, the research described here is being undertaken from the point of view of an archaeologist, currently attempting to learn the fundamentals of stone decay mechanisms [10]. Bearing in mind the possibility of the interpreter lacking a formal geological training, and the increased desire for inter-disciplinary studies and the de-compartmentalisation of approaches, the research will examine variables similar to the classification and mapping techniques of Yates *et al.* [8] and Fitzner *et al.* [9]. Thus, the distribution of various types and intensities of weathering and soiling across the stone surfaces will be interpreted and mapped digitally, and possible causes elicited. The method will be tailored towards the characteristics of the market cross, in that it will additionally explore diachronic elements, taking into account the effects of replacement parts and of changing environments for each monument.

Detailed records from close-range examination of the condition of individual stones in a representative sample of Scottish market crosses, will allow a greater understanding of the decay processes and will also enable future monitoring of decay rates. The mapped detail could also provide a visual record of conservation measures.

5. Research objectives and methodology

The main elements of the research programme are outlined as follows:

- *The history of the market cross will be discussed*

The origin, distribution, architecture and use of Scottish market crosses in the past will be subject to discussion. Comparative data from outwith Scotland will also be considered in order to allow a wider context in which to view the properties of the Scottish examples.

- *An efficient method for mapping stone surfaces will be developed, in order to record the nature and level of their decay*

Methods used by other fieldworkers to record the condition of carved stone monuments will be evaluated. A pro-forma and a data-base will be designed to store the data to be collected on market crosses. Fieldwork will be undertaken on a sample of market crosses using the developed method, and following the fieldwork the effectiveness of the method will be evaluated. The mapped data of the condition of sampled market crosses will be presented digitally.

- *A record of the current condition of a representative sample of market crosses in Scotland will be established*

A representative sample of Scottish market crosses will be selected, based upon archived data, with regard to aspects such as stone type, morphology, environmental factors, geographic location and history of intervention. Fieldwork will be undertaken to record in detail the current condition of each monument within this sample, examining a wide range of variables. An archive will be compiled consisting of drawings, photographs and collected data values, in electronic and paper format.

- *The extent to which market crosses are at risk from weathering and human agency will be assessed*

The extent of damage to the sampled monuments, and to the monument class in general, will be investigated by examining the collected data. Trends which suggest possible causes of decay will be identified. For example, the research will consider how differential erosion relates to factors such as aspect, or proximity to passing motor vehicles. The gathered data concerning present condition will be compared with any existing evidence from the past (e.g. old photographs), with a view to gauging rates of decay. The effectiveness of the conservation methods used upon market crosses will also be scrutinised.

- *Effective strategies will be proposed for the care of Scottish market crosses, based upon a risk assessment model*

A risk assessment model, constructed from the analysis of collected data, will enable future predictions to be made regarding the dynamics of the condition of this monument type in various environments. Here, the matrices, that is equations combining certain market cross characteristics and environmental conditions, which pose a potential threat to the monuments will be sought and identified. The levels and methods of management of the crosses in the past and present will be assessed. Appropriate management strategies and conservation methods will be suggested for the future, and criteria will be recommended for gauging when intervention is required for the long-term benefit of the monuments.

6. Conclusion

The ultimate aims of the project are to increase the profile of Scottish market crosses as a monument type worthy of responsible care and conservation, and to gain an increased understanding of the relationships between their condition, their environment and human intervention. It is hoped ultimately that the risk assessment model will be heeded so that the future management of such monuments might be better informed.

7. References

1. Small J. W., *Scottish Market Crosses* (Mackay E., Stirling, 1900).
2. Tittler R., *Architecture and Power: the Town Hall and the English Urban Community c.1500–1640* (Clarendon Press, Oxford, 1991).
3. Drummond J., Notice of Some Stone Crosses, with Especial Reference to the Market Crosses of Scotland. *Proceedings of the Society of Antiquaries of Scotland.* **4** (1860-1), 86-115.
4. Small J.W., Scottish Market Crosses. *Transactions of the Stirling Natural History and Archaeological Society* (1890), 51-58.
5. d'Alviella G., *Les Perrons de la Wallonie et les market-crosses de l'Ecosse.* (Hayez, Brussels, 1914).
6. Hamilton P. The Boundary Stone and the Market Cross. *Scottish Historical Review*, Oct (1914), 24-36.
7. Black W.G., *The Scots Mercat Cross: An Inquiry as to its History and Meaning* (William Hodge & Co Ltd, Glasgow and Edinburgh, 1928).
8. Yates T. Butlin R. and Houston J., *Carved Stone Decay in Scotland Methodology for Assessing the Decay of Carved Monuments for Historic Scotland.* (BRE, Watford, Jan 1997). Draft document for internal review only.
9. Fitzner B. Heinrichs K. and Kownatzki R., *Classification and Mapping of Weathering Forms.* Proceedings of the Seventh International Congress on the Deterioration and Conservation of Stone, LNEC, Lisbon, 1992, (1993) 957-968.
10. Winkler E.M., *Stone in Architecture: Properties, Durability.* 3rd edition. (Springer-Verlag, Berlin, 1994).

8. Acknowledgements

The author wishes to acknowledge the collaboration of Historic Scotland and funding from the School of Construction, Property and Surveying and the Masonry Conservation Research Group at the Robert Gordon University. The project is being undertaken as a post-graduate research degree and is due for completion in October 1999.

Stone Weathering and Atmospheric Pollution Network'97: Aspects of Stone Weathering, Decay and Conservation.
Edited by M.S. Jones & R.D. Wakefield © 1998 Imperial College Press.

CONTINGENT VALUATION COMES TO TOWN

C.N.GOMERSALL
Department of Economics and Business, Luther College,
Decorah, IA 52101, U.S.A.

Historic buildings and monuments are valued, not only for the satisfaction they may give when visited, but also for their very existence. Policymakers will want not only to recognise such "non-use" values; they will need to measure them. Contingent valuation (CV), an approach which has been used for decades in the area of natural resources, provides one such tool of measurement. In this paper, criticisms made of CV are examined and found, in some cases, either to apply with less force in the context of the built environment, or to rest on an erroneous view of CV as naive and simplistic. Only four applications of CV to the built environment could be found, and these are very briefly reviewed.

1. Introduction

Contingent valuation (CV) is one way to measure, in money terms, certain values associated with public goods[a] that cannot be established in a market; CV, one could say, measures values other methods cannot reach. In the context of the built environment, a policymaker may wish, not only to assess what the direct costs and benefits of a proposed change might be, but also to capture[b] the values which people who never visit the building or monument in question, or use it in any other way, might put on its being preserved—and then to use those values as part of a cost-benefit analysis. To describe CV in a sentence—and so risk a simplification which can lead to misunderstandings, as will be

[a] "Public goods" usually have these two characteristics: it is very hard to exclude people from receiving their benefits, and one person's enjoyment of them does not (up to the point of congestion) diminish their simultaneous enjoyment by another. In this paper, "goods" such as historic monuments and public buildings are usually considered even though, in practice, they charge admission so that they are not, strictly, "public".

[b] This somewhat martial verb is deliberate. There are other means of assessing attitudes to public goods, such as the kind of assessment, using semantic differentials, obtained by Andrew [1] for soiled and cleaned building facades in various Scottish cities; other approaches to assessment from the field of environmental aesthetics may be found in the bibliographies of Andrew and Porteous [2]. Contingent valuation, however, not only identifies attitudes, or preferences, or values; it "captures" them by applying a metric which can be used in cost-benefit analysis.

shown below—it estimates total value (including "non-use" value) by asking people how much they would be *willing to pay* (WTP) for a specific environmental improvement, or to avert a specific environmental loss[c]. Examples might be restoration of masonry on the one hand, or a decrease in the rate of its deterioration on the other.

Contingent valuation has been widely used in the area of natural resources for more than twenty years but, so far, there have been few applications to the built environment. The possibility of extending CV to this relatively new area has been noted by others such as Mohr and Schmidt [3], and this paper is, in a sense, an extension of their brief introduction to CV. Although researchers using CV in other areas have established something like a code of best practice, the technique has continued to attract certain criticisms. (One can say "continued", since many of these criticisms have been made of CV since its early days.) Anyone thinking of using CV in the context of the built environment ought, then, to examine these criticisms, and to consider if they have greater or lesser force in what is a relatively new setting. This paper suggests that, at least in some respects, there may be *fewer* objections to the use of CV in the case of the built environment than in the area of natural resources where its track record is much more extensive.

2. A review of previous contingent valuation studies of the built environment

There have so far been few contingent valuation studies of the built environment—few enough for each to be very briefly described here.

- *Valuing our cultural heritage: a contingent valuation survey* [4]

  Visitors to Nidaros Cathedral were asked how much they would be willing to pay for a reduction in damage to the building from air pollution. This cathedral is of national significance in Norway, and the authors found that even visitors (who might be expected to value it more for their "use" of it than for its mere "existence") in fact reported values related to personal use that were only 14% of their total values.

- *Valuing damage to historic buildings using a contingent market: a case study of road traffic externalities* [5]

[c] This paper uses "environment" to include the built as well as the natural one—if the distinction must be made.

Residents of Neuchatel were asked how much they would be willing to contribute to a fund set up for the maintenance of certain historic buildings in the Swiss town. The authors supposed that respondents would admire these buildings (use value) and would also be concerned about their future availability, for themselves or for their descendants (non-use values), but did not seek to distinguish between these motives.

- *Paying for heritage: what price for Durham Cathedral?* [6]

Visitors to the cathedral, for which there was no entry fee at the time, were asked how much they would be willing to pay for admission. The study was explicitly intended to value use. Willis recognised that stated willingness to pay might include a component of preservation value, but he asserted—without giving evidence—that "most [of the WTP] is likely to be motivated by WTP to gain access". Values that others, who never visit Durham Cathedral, might nevertheless place on the preservation, were not measured by this study.

- *Valuing acid deposition injuries to cultural resources* [7]

This study was commissioned by the (U.S.) National Acid Precipitation Assessment Program, who were concerned with assessing the impact of the Clean Air Act of 1990. Respondents in two major American cities were shown photographs of monuments in Washington, D.C., said to portray them in their current state and as they might be in 75 and 150 years time. Respondents were then asked how much they would be willing to pay for a coating (actually fictitious) to be put on the monuments—if the effect were to extend the period of time before the specified damage occurred.

3. Objections to CV examined in the context of the built environment

A short paper such as this, which does no more than introduce CV to those who may not previously have considered its use, can hardly summarise, let alone describe, all the debates which have attended CV over the last twenty years or more. Also, the theory behind CV draws on a great many disciplines—established ones such as economics and psychology, as well as newer ones such as environmental aesthetics—so that any short treatment may seem derisory to specialists in these related fields. The inevitable omission of this or that critical reference will raise suspicion that here, surely, contingent valuation has something to hide. There is also the risk of appearing defensive, or of setting up a straw man ("CV has its critics, but they are all wrong").

The intention of this paper is not to refute all charges made against CV (nor to mention those points in its favour which others have made) but to show that—given its track record in other fields—CV is a tool that a policymaker concerned with the built environment might usefully consider.

Although, in the case of CV, "philosophical questions are at least as important to consider as methodological questions" [8], the latter are touched on first because some, at least, stem from a common but misplaced view of CV.

3.1 Methodological objections

As examples of methodological objections to CV, consider those made by Binger, Copple and Hoffman[d] [9]: that the hypothetical nature of the case will lead to respondents' giving strategic (and usually inflated) estimates of their willingness to pay; that respondents generally lack sufficient information to value the good; that they do so without considering the effect of such payments on their own budgets or on any payments they might make for other environmental goods.

Only one of these points will be taken up here, because of the frequency with which it is taken for granted without further discussion: that respondents have an incentive to give "strategic" answers [10]. For instance, respondents might mention a lower amount than their true willingness to pay in the hope that they could enjoy the good once other people had agreed to finance it; conversely, they might respond with an unrealistically high value if they felt that there was little actual chance of their having to pay that amount. Mitchell and Carson [11] a standard reference, put forward evidence from experiments, from psychology, and from comparisons of CV findings with actual values expressed in referenda, all suggesting that strategic behaviour need not be a major problem for the analyst using CV (it *need* not be a problem, but care must still be taken). Nevertheless, the presumption—by readers—of strategic behaviour is so common that Mitchell and Carson advise all researchers using CV to address the point in their reports and show how its effects have been mitigated.

Likewise, the other objections raised by Binger, Copple and Hoffman can frequently be offset by careful survey design. Mitchell and Carson, for example, suggest ways in

[d] These objections were raised in the context of natural resource damage assessment. Since few CV studies have been carried out in the built environment, in a sense its critics have yet to catch up with it.

which these (and other) possible sources of bias might be neutralised, but they are not sanguine about such methodological issues; indeed, they continually warn that analysis using CV is necessarily complex. Contingent valuation studies do much more than ask for opinions[e]. They do not just ask "how important" the good is to the respondent—a question which might receive a facile answer. Instead, CV presents the respondent with a plausible, though hypothetical, situation analogous to a real market (or, sometimes, a hypothetical referendum). Indeed, it is a limitation of a short paper such as this that it cannot suggest the flexibility and judgment needed for useful work using CV.

The emphasis by authors such as Mitchell and Carson on nuance and on the need for exquisite care in survey design has, in turn, been characterised by Diamond and Hausman [12] as a standing admission that CV is currently flawed, but that more detailed, more complete work in the future will get it right; they also attribute to CV practitioners a belief that "some number is better than no number". Be that as it may, the opinion here ascribed to Mitchell and Carson (that CV is a difficult technique requiring considerable judgment) seems a little more accurate than that held by Binger, Copple and Hoffman[f]—the last sentence of whose article looks forward to a time when "the inertia of quick and easy valuations (such as CV) give(s) way to thoughtful and reasoned analysis." Those interested in the application of CV to the built environment might look, for example, at Morey *et al.* [7] and decide whether its analysis is "quick and easy" or "thoughtful and reasoned".

3.2 Objections to the subjective nature of CV

Contingent valuation assumes that value is not innate in a good, but is attributed to it by the valuer, and this characteristic gives rise to the two kinds of criticism introduced in this section: that certain goods have intrinsic value or, alternatively, that respondents—or other humans—have rights to those goods such that "willingness to pay" is not an appropriate concept (intrinsic value and the question of rights are often confused but as

---

[e] Laing and Urquhart, in this volume, express a common view when they write, in describing CV *obiter dicta*, that "respondents are asked *simply* to state" their WTP (emphasis added).

[f] It must be pointed out that Binger, Copple and Hoffman do carry out their own contingent valuation study, which is more than most critics of CV are prepared to do.

stated below and as may be more fully explored in the cited literature, they are in fact conceptually distinct).

It is sometimes said that CV is utilitarian but, in principle, this need not be the case: for the respondent, the good might indeed have innate[g] value. Contingent valuation does not see a particular monument as embodying "conservation value" or "heritage value"— but would not deny that respondents might *attribute* such values to the monument in question [13]. To call such a response utilitarian—to suppose, for example, that "heritage value" provides utility to the respondent—is to reason tautologically [14].

Whether or not it is utilitarian, however, CV does indeed rest in the preferences (or opinions, or values; again, the wording can be treacherous) of the respondent. Although this principle has its roots in a hundred years of economic thought, it is interestingly consistent with postmodern ideas in literary and other artistic criticism—whose recent application to cultural resources is touched on, for example, by Cahn [15], writing on the Romanesque Frieze at Lincoln Cathedral. This section, then, introduces complaints made about the subjective nature of contingent valuation.

♦ *Intrinsic value*

Intrinsic value can be interpreted in many ways. This section follows Regan [16]. Goods may be said to have intrinsic value as ends in themselves and, if so, it is inappropriate to ask whether a respondent is willing to pay for a change in the quantity or condition of these goods. Those who hold such views are required, however, to specify *which* goods have this value (it is a categorical value; either a good has intrinsic value, or it does not). Kant finds this value in the rationally autonomous individual; others have found it in non-human beings of various kinds. Regan (writing, admittedly, in a volume devoted to the intrinsic value of natural resources) mentions no case in which artefacts are held to be ends-in-themselves. Since not everything can be an end-in-itself (some goods must have instrumental value), it is far from clear that artefacts in the built environment could have intrinsic value of this kind.

It may, on the contrary, be maintained that certain states of affairs possess intrinsic value[h]. Examples might include a particular sunset, which could be said to have intrinsic value whether or not there was anyone watching to appreciate it. It is hard to imagine

[g] Terminology in this area is by no means agreed. This paper uses "intrinsic value", "inherent value", and "innate value", with a certain degree of ambiguity.

[h] Interestingly, Cox [17] attributes this view, not to critics of CV as does Regan [16] but to some of its *practitioners*.

cases of this kind in the context of the built environment. One might suppose, for example, that it is "intrinsically" valuable that buildings be extant from the entire history of a community (a state-of-affairs), but it is harder to suppose that this value would remain if there were no one to live in the community—or even to visit its remains as tourists.

Even if it were possible to view states of the built environment as intrinsically valuable in this sense, such values do not necessarily imply any obligations on our part towards the good in question [17]. In policy terms, one is still free to consider, in a cost-benefit framework, the subjective values of community members (extrapolated from a sample of respondents).

♦ *Rights-based objections*

Subjects may feel that rights are vested, either (i) in the good itself, especially if it is a natural resource; (ii) in other humans, including future generations; or (iii) in the respondent. The first two of these views might provide grounds to protest against the subjective nature of a CV questionnaire. (The third view of rights, vesting them in the respondent, is treated separately below.) It is legitimate to ask questions about rights even if one believes, with MacIntyre [18], that they may be "moral fictions". The attitude of the respondent is, after all, the relevant issue.

♦ *Rights of the resources being valued*

As noted above (i) only (some) *natural* resources have ever been considered to have status as ends-in-themselves; and (ii) in any case, intrinsic value and moral standing are distinct concepts. Nevertheless, some consideration ought to be given here to the "rights" of objects in the built environment if, for example, respondents feel that they should be protected through legislation rather than financed through the tax or fee identified in a CV study[i].

Hanley and Milne [19] have sought to estimate the degree to which respondents in CV studies of natural resources, basing their values on the concept of rights, give answers which vitiate the study as a whole. In their survey of the empirical literature, they find that 24% of respondents in one study, and 23% in another, give answers that imply a rights-based approach to the resources in question. In their own empirical work, designed explicitly to estimate these percentages, they find that, of 30 zero bids out of 97 responses, 14 "bid zero for reasons other than a zero value; most common of these reasons were that

[i] Even if legislation is passed someone must pay for restoration or, in general, for mitigation of damages. These costs may indeed result in payment of taxes or fees by the respondent. If, however, the respondent feels (albeit myopically) that "legislation" is the answer, then he or she may give a zero protest bid.

the environment should be protected by law. These respondents (14% of the sample) could be described as having a rights-based belief in the environment".

- *Rights of future generations (and of others)*

Although historic monuments may not have rights, they do—unlike natural resources—have builders who may, in the view of some, have rights to their work for ever after. One gets the sense, for example, in this passage from Ruskin, that he might not have approved of contingent valuation:

"... it is again no question of expediency or feeling that we shall preserve the buildings of past time or not. *We have no right whatsoever to touch them*. They are not ours. They belong partly to those who built them, partly to all the generations of mankind who are to follow us. The dead still have their right in them..." (quoted by Lasko [20]; emphasis in original). Here, it seems, are some "stakeholders" in the built environment (and not in the natural), in contrast with the *lack* of rights assigned to the cultural (again, as opposed to natural) goods in themselves. However, Ruskin not only assigns rights to builders and to posterity, he also denies rights to his own contemporaries (and, by extension, to ours). Yet much of the built environment explicitly possesses a public aspect [15]; whether or not the builder intends the spectator to be instructed, edified, impressed, or moved by his work, that the spectator should *respond* is inherent in the nature of most of the buildings and monuments with which policymakers must deal. It is thus legitimate, and not just a play on words, to recruit that spectator as a respondent in a CV survey.

Contemporary respondents may also be thought of as acting by proxy for future generations, there being no other way to capture what will again, in due course, be a *response* to the monument on the part of those not yet born (although this view is, of course, itself contentious).

There remains the question of whether the rights of the builder (if they are conceded) are to be lexicographically preferred to the values given the monument by generations of spectators, today or yet to come. One cannot suppose that the spectators may themselves act as proxies for the builder, since the spectator of a monument—particularly of a monument from antiquity—is unlikely to be the spectator for whom it was initially designed.

Suppose, as a case in point, that Shelley's statue of Ozymandias has been found, and a decision is to be made on its treatment. It would be quite impossible to respect the sculptor's supposed wishes and rights: no one today or in the future would be awestruck by Ozymandias' power, especially given the statue's dilapidated state. Ruskin's view is, to put it no more strongly, not always suitable as a basis for policy.

♦ *Respondents' rights*

Subjects might feel that, because of their rights to enjoy the good in its present state, they should not have to pay anything at all to delay its deterioration. Zero responses of this kind must, of course, be distinguished from those where the respondent actually places no value on the resource, or where the respondent is financially constrained. This distinction can sometimes be made through a follow-up question which asks those respondents who give zero values for their motives in doing so.

In the CV studies of the built environment identified above, the number of zero responses were as follows:

1. Navrud [4]: Information on protest bids was not available for this paper.
2. Grosclaude and Soguel [5]: Of the 200 respondents, 86 reported zero WTP. Follow-up questions determined that, of these, 22 were genuinely indifferent to the goods being valued; 16 were not willing to pay because of financial constraints; and 48 (24%) were described as "free-riders" whose actual WTP was positive despite their zero bids. "According to them they already pay enough taxes and the state or the road users should be charged."
3. Willis [6]: "Only 5% of those approached refused to participate", but the study does not disclose (i) whether people were told the nature of the study before they were asked to take part, and (ii) the percentage, of those responding, who gave zero bids for protest reasons.
4. Morey *et al.* [7]: Subjects did not know, before the interview began, exactly what it would cover. Nevertheless, of the 237 respondents providing answers to a WTP question in payment card format[j], only eighteen gave zero values. Of these, fourteen had said (in response to previous attitudinal questions) that they did not care about preserving the monuments; two were ambiguous, and only two stated that their zero bids were a protest in the sense being used here.

The evidence from these studies is sparse, although the study by Morey *et al.* [7]. does suggests a very low level of rights-based protest. None of the studies identifies zero bidding with the perspective, reported by Hanley and Milne [19] that the good should be

[j] This format was actually secondary in the study; more prominent was a paired-choice format. Protest bids from the payment card format are used here as being more conventional.

protected by law. Moreover, the study by Grosclaude and Soguel [5] points up a difficulty in attributing zero bids to a basis in rights of the respondent. These authors explicitly state that "according to" 24% of their respondents, they "already pay enough taxes"—yet the authors also describe these respondents as "free-riders". Free-riding is another form of strategic behaviour, not of rights-based protest. Mitchell and Carson [11] point out that it is most likely to be found where respondents feel that the good will be provided whatever they say, but that, on the other hand, their share of the cost may be determined in part by any amount they mention—both circumstances that may, perhaps, be found in Grosclaude and Soguel's scenario. Whether or not this is so for their study, it is hard in general to ascribe a statement that "someone else should pay" to a rights-based perspective when it may instead describe strategic behaviour[k].

3.3 CV and socio-politics

Some find it wrong to use CV in decisions that should call for resolution through the political process [21]. They stress that public goods should be valued through a deliberative process, rather than atomistically through a device such as a CV study. In so far as policymaking for the built environment has to do with such issues as health, this criticism of CV may have some force (some might say that even aesthetic concerns have a social aspect). On the other hand, it may be a counsel of perfection to insist that such questions can be settled through debate. In practice, participation in such debates may be limited not only in numbers but also in character: argument may be limited to the exchange of expert opinion. Mitchell and Carson [11], for instance, actually find CV to be commendably democratic. It gives equal value to the views of every respondent[l], whether expert or not—and some of these views may indeed be formed by debate.

[k] Although less plausibly in view of their wording, the protest bids reported by Hanley and Milne [19] could also be interpreted as disguised "free riding" rather than rights-based objections.

[l] The usual difficulties in sampling from more transient (and some other) sections of the population remain, as always, to be dealt with.

3.4 Objections based on economic orthodoxy

There is no doubt that many economists are dubious about CV because it departs from the assumption that preferences are revealed only in actual (and highly stylised) exchange. Mitchell and Carson [11] do their best to show that CV is consistent with other tenets of mainstream economics, but perhaps this is an essentially contested issue on which it is impossible to reach consensus through debate. On the other hand, surveys in other branches of economics are starting to make their presence felt. One influential proponent has been Alan Blinder, until recently Vice-Chairman of the (U.S.) Federal Reserve Board. In any event, contingent valuation has received some recognition in the (U.S.) courts, if not from all in the economics profession.

4. What does contingent valuation measure?

A number of contingent valuation studies have sought to measure the overall value given by respondents to a good, and then to "back out" some measure of *direct* use, preferably determined by market behaviour. The (U.S.) Department of the Interior recommended this approach, as pointed out with some scepticism by Binger, Copple and Hoffman [9]. "As a practical matter, it will usually not be necessary to categorize particular nonuse values during a natural resource damage assessment....although the Department recognizes that the definition is somewhat circular, the Department believes it is appropriate to define 'nonuse value' as the difference between [total] value and use value." [22]

Just what it is that remains after subtracting, say, market values from a measure of total value has sometimes given rise to confusion and disagreement. In the case of the built environment, one might be tempted—despite the recommendations of the U.S. Department of the Interior recommendations mentioned above—to break down this residual into the values Porteous [2] puts forward in connection with "valued landscapes": values that could be attributed, not only to aesthetics, but also to "familiarity, comfortableness, affection, and attachment".

In practice, however, people may be unable to distinguish, through introspection, between the various motives they might have for attributing value to a particular good [11]. Studies which ask respondents to do so—e.g., Stevens *et al.* [23], who report a division of total willingness to pay into use value (7%), bequest value (34%) and intrinsic value (48%)—run the risk of a factitious precision. At best, a market-derived use value can be subtracted out, leaving one with an undifferentiated residual analogous—and the analogy is not entirely a happy one—to total factor productivity[m] in standard theories of economic growth.

If it is impossible, in practice, for respondents to distinguish their various motives through introspection, it is nevertheless most important for analysts to associate willingness to pay, as best they can, with concepts such as "existence value", "heritage value" and the like. After all, it is vital that respondents be invited to consider *all* possible sources of value they might have [11] and, without the terms' necessarily being used, such concepts as existence value need to be explained to respondents in order for them to reveal more accurately their true willingness to pay for the public good.

5. Conclusion

Contingent valuation is all too easily caricatured as a simple opinion survey and, perhaps as a result of this apparent simplicity, as possibly "one more historic relic in the museum of junk science" [9]. Alternatively, it has sometimes been written off as walking on holy ground—as trespassing in areas where measurement of any kind is inappropriate. These assertions have been considered above, if only in brief.

Theorists who find CV defensible, moreover, and practitioners who have used it over the past twenty or thirty years, do not necessarily take a reductionist view; they do not usually suggest that CV is a *sufficient* guide to decision making in a policy context. Instead, they are more likely to recommend that CV be used as one strand in the formation of policy.

For it is likely that the built environment, relatively familiar as it is to most citizens, will be a public good on which ordinary people do have opinions, preferences, and indeed values. Omission of these values *within* cost-benefit analysis cannot be desirable, and contingent valuation, far from being a blunt instrument as sometimes portrayed, offers one way—an intricate and even delicate way—to consider such values in a systematic fashion.

---

[m] The two main factors usually held to influence economic growth are labour and capital. Changes in the productivity of these two factors, however, are not sufficient to account for overall growth; a residual remains that is known as total factor productivity (TFP). It has been hard for economists to explain TFP, though growth theory is moving away from the simplistic model introduced here for purposes of analogy.

6. References

1. Andrew C.A., *An Investigation into the Aesthetic and Psychological Effects of the Soiling and Cleaning of Building Facades* Ph.D. dissertation (The Robert Gordon University, 1993).
2. Porteous J.D., *Environmental Aesthetics* (Routledge, London, 1996).
3. Mohr E. and Schmidt J., Aspects of Economic Valuation of Cultural Heritage, In *Saving Our Architectural Heritage: The Conservation of Historic Stone Structures* ed. Baer N.S. & Snethlage R. (John Wiley, Chichester, 1997) 333–348.
4. Navrud S., ed. *Pricing the European Environment* (Oxford University Press, London, 1992).
5. Grosclaude P. and Soguel N.C., Valuing Damage to Historic Buildings Using a Contingent Market: A Case Study of Road Traffic Externalities *Jnl of Environmental Planning and Management*, **37** no.3 (1994) 279–287.
6. Willis K.G., Paying for Heritage: What Price for Durham Cathedral? *Jnl of Environmental Planning and Management*, **37** no.3 (1994) 267–278.
7. Morey E. Rossmann K. Chestnut L. Ragland S., *Valuing Acid Deposition Injuries to Cultural Resources* on-line available @ http://spot.Colorado.EDU/~morey/monument/ toc. htm (no date)
8. Mazzotta M.J. and Kline J., Environmental Philosophy and the Concept of Nonuse Value. *Land Economics*, **71** no.2 (1995) 244–249.
9. Binger B.R. Copple R.F. Hoffman E., The Use of Contingent Valuation Methodology in Natural Resource Damage Assessments: Legal Fact and Economic Fiction *Northwestern University Law Review*, **89** no.3 (1995).
10. Schubert U., The Preservation of Stone Monuments in the Maze of Complex Urban Development Policy, eds Baer N.S. & Snethlage R. in *Saving Our Architectural Heritage: The Conservation of Historic Stone Structures* (John Wiley, Chichester, 1997) 315–332.
11. Mitchell R. and Carson R., *Using Surveys to Value Public Goods* (Resources for the Future, Washington, 1989).
12. Diamond P.A. and Hausman J.A., Contingent Valuation: Is Some Number Better Than No Number? *Jnl of Economic Perspectives*, **8** no.4 (1994) 45–64.
13. O'Neill J., The Varieties of Intrinsic Value *The Monist*, **75** no.2 (1992), 119–137.
14. Robinson R., *Definition* (Oxford, Oxford University Press, 1950).

15. Cahn W., ed. Romanesque Sculpture and the Spectator, in Kahn D. *The Romanesque Frieze and its Spectator* (Harvey Miller Publishers, London, 1992), 45–60.
16. Regan T., Does Environmental Ethics Rest on a Mistake? *The Monist*, **75** no.2 (1992) 161–182.
17. Cox J.R., The Relations Between Preservation Value and Existence Value. In *Valuing Nature?* ed. Foster J. (Routledge, London, 1997) 103–118.
18. MacIntyre A., *After Virtue* (Duckworth, London, 1985).
19. Hanley N. and Milne J., Ethical Beliefs and Behaviour in Contingent Valuation Surveys. *Journal of Environmental Planning and Management*, **39** no.2 (1996), 255–272.
20. Lasko P., The Principles of Restoration, ed Kahn D. in *The Romanesque Frieze and its Spectator* (Harvey Miller Publishers, London, 1992) 143–162.
21. Jacobs M., Environmental Valuation, Deliberative Democracy and Public Decision-Making Institutions, ed. Foster J. in *Valuing Nature?* (Routledge, London, 1997), 211–231.
22. Department of the Interior, *Natural Resource Damage Assessments; Proposed Rule*, in Federal Register **43** CFR Part 11 (July 22, 1993).
23. Stevens T.H. Echeverria J. Glass R.J. Hager T. More T.A., Measuring the Existence Value of Wildlife: What do CVM Estimates Really Show? *Land Economics*, **67** no.4 (1991) 390–400.

7. Acknowledgements

The author much appreciates the comments made by two anonymous referees.

Stone Weathering and Atmospheric Pollution Network'97: Aspects of Stone Weathering, Decay and Conservation.
Edited by M.S. Jones & R.D. Wakefield © 1998 Imperial College Press.

CLEANING OF STONE BUILDINGS: THE APPLICABILITY OF ESTABLISHED VALUE ASSESSMENT METHODOLOGIES

R.A. LAING AND D. URQUHART
Masonry Conservation Research Group
Faculty of Design, The Robert Gordon University,
Garthdee Road, Aberdeen AB10 7QB, Scotland, UK.
E mail r.laing@rgu.ac.uk

The measurement of value obtainable through conservation of the built environment requires a number of approaches to be used in association with one another. The value system surrounding the built environment is complex to the extent that any one approach could be accused of failing to represent the holistic effects taking place. Therefore, this research aimed to establish links between value and conservation, and in so doing has identified methodologies which might be used in the future to predict changes in value prior to the commencement of work. The cleaning of stone buildings has been used as the study topic throughout.

1. Introduction

The practice of stone cleaning has a considerable effect on the value system associated with the built environment. That value system consists of a number of inter-related variables concerning the aesthetic, financial, environmental and heritage aspects of the built environment, and like any value system is open to influence and pressure. Stone cleaning is capable of having an impact on some or all of these constituent parts, and it is the identification and assessment of these influences and changes which will be discussed here. Indeed, the wide application of cleaning methods and their continued use must be seen as being driven by anticipated changes which will result within that value system (e.g. changes in the aesthetic or financial value). Therefore, any guidelines or advice produced to inform parties interested in the effects of stone cleaning should be linked to the value system, in order that such advice is adequately linked to the mechanisms affecting value change. This paper outlines a possible approach to value assessment that incorporates the effects of stone cleaning, and one which has been applied through the pursuance of a Ph.D. research project.

2. Value system associated with stone cleaning

Methods of value assessment, whilst allowing certain aspects to be defined and measured, present problems where the representation of a holistic value system is concerned. That is, the assessment of facets within that value system lends itself to the use of relatively simple methodologies, where the adequate measurement and subsequent comparison between likely outcomes is possible. Other aspects of the value system, however, exist in situations where the aspects of value to be assessed are relatively difficult to measure, allowing little direct comparison between variables.

Figure 1 presents the major variables which should be regarded as together representing the overall value system, with suitable groupings for assessment purposes suggested. The diagram represents the hypothesis that overall value can only be understood after a thorough examination of it constituent parts has been completed. Financial and environmental value can be objectively assessed using established methodologies, whereas the assessment of heritage value requires the recognition of more subjective criteria.

In addition, although parts of the overall value can be extracted for analysis, the overall system is holistic in nature, and must be understood to be more than simply a sum of its parts.

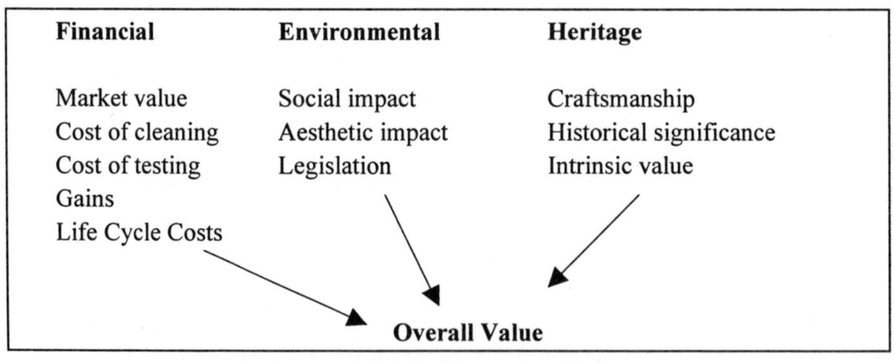

Figure 1. Variables to be considered in a value assessment of stone cleaning

In order that the outcome of any value assessment can be regarded as having relevance within any possible future application, it is first necessary that the methods of assessment to be followed are clear, make a realistic use of data and that the range and depth of the information required is realistic. For these reasons a number of established value assessment techniques have been used, during the PhD study, with the aim of ensuring that the consequences for value attributable to cleaning can be adequately assessed prior to any future cleaning interventions.

2.1 Financial value

The modelling of financial value has been practised within the built environment in some form or another for well over a century. This has led to the development of practical and reliable methodologies which may be applied by the practitioner. Such methodologies rely on the cost data to be used being both reliable in terms of its magnitude and predictable in its frequency.

It was necessary for the purposes of the current study to determine initially the types of financial value which might be affected by stone cleaning. The major variables, for which the individual effect of each on financial value can currently be predicted, are banded broadly as follows:

i. initial costs of cleaning
ii. associated initial costs
iii. subsequent maintenance and repair costs
iv any financial gains which might accrue from cleaning

It is possible to model the financial effect of variables i, ii and iii for predictive purposes due to the availability of cost/benefit data. A degree of uncertainty regarding the prediction of maintenance and repair frequency is being currently addressed by the Masonry Conservation Research Group, Aberdeen. A study was completed as part of the research project regarding variable iv., in order that a conceptual model of financial change could be compiled.

Points i. and ii. include factors such as the cost of cleaning itself, the cost of any temporary works required (e.g. scaffolding, protection of public), the cost of any testing required prior to the commencement of cleaning and costs associated with re-pointing or immediately necessary repair work. Within these areas, although it is possible to generalise in terms of the variables to be considered, the costs themselves will depend upon the size of the facade, the type of cleaning method proposed and the condition of the stone itself. In addition, it is usual for cleaning work to be implemented alongside other construction work, in which case the cost of scaffolding can be apportioned between work elements, theoretically at least reducing the direct cost of cleaning. Once each variable has been identified and costs arrived at, the production of an estimate becomes relatively simple.

However, whereas points i. and ii. can be estimated using estimation techniques commonly used within the construction and building conservation industries (but still depending on reliable data being available), subsequent maintenance and repair costs present a number of problems. Perhaps the most pressing of these is the need for reliable data concerning the likely frequency and magnitude of any costs incurred over the residual life span [1,2]. As with any prediction, a certain degree of uncertainty exists. This should be included as part of the overall model, after such uncertainty has been reduced as far as possible. For the present study, the life cycle has been taken to refer to the life span of the building, as opposed to the time at which it is occupied or let by the present owner.

The possible gains referred to in point iv. can again be grouped under two main headings, these being:

I. gains associated with the property market value
II. financial benefits to non residents/owners within the wider community

A major study conducted as part of the study focused on the first of these points, and resulted in the production of likely percentage changes in the market selling price of properties due to stone cleaning. These indicators can be employed as part of future predictions [3]. The second point refers to wider benefits, including impact upon retail, commercial and tourism revenues. This study found, when examining the effects of stone cleaning on property market selling/letting prices that stone cleaning was perceived within property markets as leading to improvements in the marketability of buildings for sale or lease. Were such improvements to be felt within the tourism market, for example, benefits to the wider community might well compare with the financial costs of cleaning outlined above. However, the extent to which cleaning when taken in isolation would be likely to result in significant macro-economic gains is open to debate, and is difficult to assess in relation to one particular building or cleaning contract.

For each financial cost or benefit likely to be incurred over the life cycle of the building, as opposed to immediately after stone cleaning, it is essential that a model of the likely magnitude of such life cycle costs is developed. At the present time, detailed information regarding the frequency of maintenance work associated with cleaning is unavailable, although the costs of the maintenance itself can be estimated using traditional estimation procedures. The frequency is strongly dependant on a number of factors including the pre-cleaning condition of the stone itself, the cleaning method used and the environment in which a building is located. For any particular case, it is possible to estimate ranges of likely frequency based upon expert opinion alone which should suffice for budgetary purposes. For more detailed and reliable estimates, however, work exploring the longer term effects of cleaning, such as that currently being undertaken by the Masonry Conservation Research Group at The Robert Gordon University in Aberdeen, will aid in improving such reliability. It has been found that existing methods of cost estimation are capable of producing a model which will allow the practitioner to estimate to a suitable degree the likely effect of cleaning on financial value. Where it is not possible to estimate the exact frequency or extent of work which might be required at some point during the residual life span, use of the Monte Carlo method of statistical simulation [4,5] allows for the incorporation of expert judgement and the recognition of uncertainty within the model, allowing the consideration of a wider range of factors within the calculation.

By combining a number of financial variables, where not all of the costs and benefits will necessarily be incurred or accrued by single parties, it is possible to move towards a more holistic assessment, and one which is easier to compare with those other value groupings, discussed below.

2.2 Environmental value

The concept of environmental value encompasses many aspects of overall value which cannot be assessed directly through an analysis of financial transactions, but which represent important aspects of overall value. Both "use" and "non-use" values are included (meaning both value derived through direct interaction with a building or group of buildings, and value derived solely through the existence of a building or streetscape).

A previous study [6] illustrated that the degree of soiling present on stone buildings was a criterion by which perceived judgements about those buildings are made, with that impact existing within a wider aesthetic framework [7]. Large scale cleaning programmes have made a great impact on many cities in the UK, and the full effects of the changes in terms of value cannot be reflected fully by analysing solely the financial effects described. The effect of stone cleaning felt by individuals not directly affected by or influencing the financial markets associated with a building should not and cannot be overlooked. Whilst the financial effects described could be termed "use values", "non-use values" experienced by the wider community (usually non occupants or owners) form an important part of the overall value system. Such values would include feelings of social well-being, the appreciation of the built environment as an adequate setting for society and the importance of aesthetic value within the living environment. Indeed, a deeper understanding of the popularity of stone cleaning is attainable through the analysis of such values, where any nuances of perception and attitude not obvious from observation of the financial system should become evident (e.g. knowledge and opinions of the individual, distance of site from home, current ownership of buildings).

A number of methods have been developed in the past to allow the assessment of environmental value, each of which is suited to certain types of situation [7]. The topic of stone cleaning and stone conservation is best served by the contingent valuation method [8] where, due to the absence of a "market" from which the environmental value can be extrapolated, a hypothetical market is established. This would normally take the form of a description of a situation in which a project would either proceed or otherwise, based on financial input from the respondent group. Respondents are asked to state the amount of money they would personally be willing to contribute towards the implementation of an environmental change (in the case of cleaning, to ensure for example that a certain building or area be cleaned) [9-13]. For the current study, interviews were carried out at locations where only some of the buildings had been cleaned. In this way it was possible to illustrate some effects of cleaning whilst avoiding problems or criticism due to the use of inappropriate visual aids (e.g. photographs, sketches, computer manipulated images).

Data regarding the background of the individual (age, sex, home, education, prior knowledge and opinions, occupation, salary) is collected to allow the identification of any significant relationship between social indicators and the value held for the results of stone cleaning.

A potential problem which must be addressed by any study of environmental value associated with stone cleaning concerns the manner in which the effects of cleaning are manifested. Whereas initial colour change brought about by the cleaning may well result in perceived aesthetic benefits, any subsequent problems regarding the longer term stone condition, or indeed reductions in aesthetic value due to re-soiling, will be considered by most respondents. Therefore, a complexity of response caused by the presence of both positive and negative effects might result in an aggregation within their bid, with this complexity masked by the apparently exact nature of the bid itself. Therefore, it should be reasonably expected that the social indicators for a respondent sample will be as valuable, if not more so, to the decision maker when attempting to predict the likely effects of a cleaning contract, prior to the application of works.

The environmental values associated with cleaning, whilst not reflected fully in existing financial markets, form an essential part of the overall value system. Although a model of such values would not be directly compatible with the model structure described for financial value, each takes on a greater significance when considered alongside the other. The use of such models by the decision maker allow for a greater importance to be placed on any resolutions.

2.3 Heritage value

Concern has been raised for a number of years regarding the potential effects of stone cleaning, both in the long and short terms. Indeed, in Scotland it is the case that the cleaning of listed properties requires planning permission and listed building consent due to its status as an alteration. Such protection being extended to listed buildings reflects a feeling that such properties, where their heritage value should be protected, might be damaged to some degree in the longer term, leading to a subsequent loss of heritage value. An essential question which must be addressed concerns the location of that heritage value, the extent to which it might be affected by adverse cleaning results and how resulting changes in heritage value might be assessed [14]. Stone carving detail might be said to contribute to heritage value where: the aesthetic impact created were notable in relation to that of other buildings; where the craftsmanship involved had become obsolete (or was of an unusually high standard); or where the carving work in some sense formed an important historical document. Were stone carving work on the facade deemed to contribute substantially to heritage value in any or these senses, then any damage to that carving would result in a reduction in heritage value. Were the heritage value dependant on factors largely independent of the condition of the stone, the effects of cleaning might be less critical.

Whatever the situation may be for any given case, however, the results of any studies concerning financial and environmental value should ultimately refer back to the heritage status and value of the building itself. Where the projected residual life span of a building extends for a century or more, the financial costs incurred over the next few years might well seem rather insignificant before very long.

3. Conclusions

This paper has aimed to discuss the consequences of stone cleaning for value, and the applicability of established value assessment methodologies. It can be seen that a number of discrete methods are available to investigate aspects of the overall value system, which if used together provide a route towards closer understanding of the overall, holistic, value framework. Decisions made regarding the continued implementation of stone cleaning rely heavily upon factors shown to be rooted deeply within the associated value system. The results of research into stone cleaning can be effectively communicated through the use of value prediction models, as all effects of cleaning will be reflected the value system. Aspects of the overall value system will be of importance to any party interested in the future application of stone cleaning, or the effects of past work, thus providing a platform for the assessment and presentation of the effects of stone cleaning.

4. References

1. Ashworth A., Life Cycle Costing: predicting the unknown. *Building Engineer* (April 1996) 18-20.
2. Ince R., Life-cycle costing. *Construction* (September, 1992) 30-34.
3. Laing R. and Urquhart D.C., Stone cleaning and its effect on property market selling price. *Journal of Property Research*, (in press).
4. Chau K.W., Monte Carlo simulation of construction costs using subjective data. *Construction Management and Economics*, **13**, (1995) 369-383.
5. Wall D., Distributions and correlations in Monte Carlo simulation. *Construction Management and Economics.* **15**, (1997), 241-258.
6. Webster R. Andrew C.A. MacDonald J. Thomson B. Tonge K. Urquhart D.C. and Young, M.E., Masonry Conservation Research Group, RGIT, Aberdeen, *Stone Cleaning in Scotland, Research Commission investigating the effects of cleaning of sandstone*, Report to Historic Scotland and Scottish Enterprise (1991).
7. Nasar J. Urban Design Aesthetics, the evaluative qualities of building exteriors. *Environment and Behaviour*. **26** (3) (1994) 377-401.
8. Mohr E. and Schmidt J., Aspects of Economic Valuation of Cultural Heritage. In *Saving Our Architectural Heritage: The Conservation of Historic Stone Structures,* ed. Baer N.S. and Snethlage R., (John Wiley and Sons Ltd, 1997) 349-370.
9. Hanley N., Using contingent valuation to value environmental improvements. *Applied Economics*. **20**, (1988) 541-549.
10. Hanley N., Valuation of Environmental Effects, *ESU* Research Paper No.**22**, Industry Department for Scotland, Scottish Development Agency (1990).
11. National Oceanic and Atmospheric Administration, Report of the NOAA Panel on Contingent Valuation. *Federal Register*. **58**, No. 10, Jan 11 (1993), 4602-4614.
12. Pearce D., (HMSO, London. 1991). Policy Appraisal and the Environment.

13. Willis K.G. and Garrod G.D., Landscape Values a Contingent Valuation Approach and Case Study of the Yorkshire Dales National Park, Countryside Change Initiative, Working Paper **21**, Department of Agricultural Economics and Food Marketing, University of Newcastle Upon Tyne (1991).
14. Snickars F., How to assess and assert the value of the cultural heritage in planning negotiations. In *Saving Our Architectural Heritage: The Conservation of Historic Stone Structures,* ed. Baer N.S. and Snethlage R. (John Wiley and Sons Ltd, 1997), 349-370.

5. Acknowledgements

The project outlined in this paper was jointly funded by the Technical Conservation, Research and Education Division of Historic Scotland, and The Robert Gordon University, Aberdeen.

Stone Weathering and Atmospheric Pollution Network'97: Aspects of Stone Weathering, Decay and Conservation.
Edited by M.S. Jones & R.D. Wakefield © 1998 Imperial College Press.

STONE CONSERVATION WITHIN HISTORIC SCOTLAND FROM 1954 TO THE PRESENT DAY

N. BOYES
Historic Scotland TCRE 2, 7 South Gyle Crescent
Edinburgh EH12 9EB

1. Introduction

Historic Scotland's Conservation Centre is part of a broad ranging Technical, Conservation Research and Education (TCRE) division and provides for the Agency the unique resource of practical expertise in conservation. The work is primarily concerned with Historic Scotland's own monuments of which there are 330, and is determined on a seasonal basis whereby work is done on site during the summer. Subjects are brought back to our studio facilities in Edinburgh for conservation during the winter months as the Scottish climate is not conducive to many outdoor conservation processes. Conservators within the Centre provide both skilled hands-on work as well as advisory and analytical services. Historic Scotland/TCRE are also in a position to provide these services to outside bodies, private clients and grant recipients as and when in-house work programmes permit. Based at two sites in Edinburgh, Stenhouse and South Gyle, the Historic Scotland Conservation Centre (TCRE2) has amassed a wealth of experience from the long history of working on Scottish Buildings and materials and subjects. The Centre's expertise comes from tackling conservation problems that arise more particularly in Scotland.

Within the Conservation Centre, staff specialise in a number of conservation disciplines: the conservation of wall and Scottish Renaissance ceiling paintings (termed structural paintings) where skill and experience have been developed over many years in the treatment of painted ceilings; the repair and conservation of easel paintings and frames are undertaken from Historic Scotland's own estate and outside bodies such as The National Trust for Scotland; the conservation of heraldic imagery; advisory environmental monitoring and analytical services. Preventative conservation is a feature of any conservation discipline and is undertaken by careful monitoring of conditions and strategic advice on handling environmental control to ensure the survival of artefacts, architectural detail and buildings with the minimum degree of intervention.

Conservators within Historic Scotland Conservation Centre also undertake the replication of subjects by the moulding and casting of monuments and carved ornament
where necessary. This is necessary upon occasion when an original subject is situated within a hostile environment and its relocation is fundamental to its ongoing care. In order to retain cosmetic integrity of the monument a mould will be taken and replicas will be made often in fibre glass and polyester resin to replace the original subject.

Perhaps due to the number and variety of stone subjects throughout Historic Scotland's estate, stone conservation and the conservation of decorated plaster work are perhaps the dominant disciplines within the Conservation Centre. The decay mechanisms encountered throughout the estate and conservation problems are as various as the subjects themselves, such as the delaminating monoliths within the Ring of Brogar on Orkney, problems associated with bird guano, organic growth, soiling by atmospheric pollution and the formation of sulphation skins for example.

The Centre plays a role in setting and raising standards in conservation within Scotland by contributing to TCRE's research and publication activities. The Centre also hosts training courses of both in-house and other personnel in conjunction with the Scottish Conservation Bureau and provides placement training for Bureau Fellows and students from other conservation courses.

2. Case Study: Stirling Mercat Cross

Illustrated within Figure 1 is a typical example of a stone conservation project conducted for an external client of the Historic Scotland Conservation Centre. This is the Mercat Cross of Stirling which was erected in its present position in Broad Street, Stirling in 1995, during part of landscaping works to create a pedestrian area within this part of Stirling. Prior to that the subject was situated just a number of feet away from where it had been erected in 1891. The three components visible are the octagonal shaft, decorated capital and unicorn finial. It is known that the finial had formed part of an older monument previously located within a niche above the tolbooth door also within Broad Street. It is thought that the unicorn finial known affectionately by the locals and conservators as "*Puggy*" is believed to date from the early 17th century. Documentary evidence exists from 1617 when in anticipation of the visit of James I and VI, the town treasurer was ordained to buy some leaves of gold to guild His Majesty's Arms of the Cross and Tolbooth. In 1792 the cross had allegedly become ruinous and was an obstruction to increasing traffic levels. Application was made to the Court of Session and permission granted for its removal. A small stone was erected at the site and deemed to be the Cross of Stirling for official purposes. The "*Puggy*" was placed in the niche described previously.

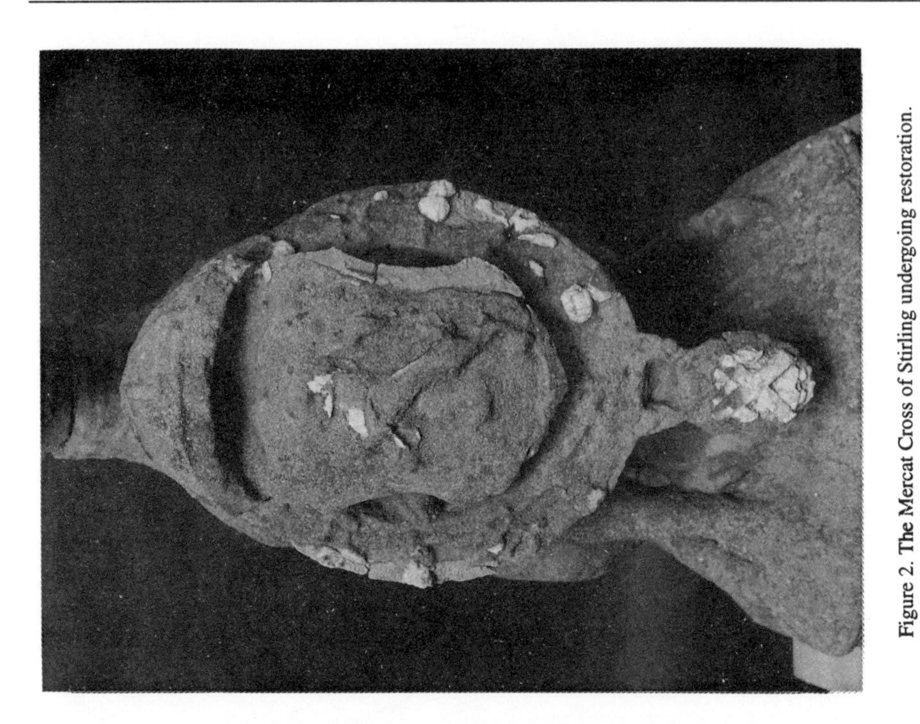

Figure 2. The Mercat Cross of Stirling undergoing restoration.

Figure 1. The Mercat Cross of Stirling.

In 1829 a letter in the local Stirling Press stated that *"Puggy's"* close proximity to local schools made it a target for stones and old shoes and that a few days before, its head had been knocked off. Subsequently, in 1830 it is documented that *"Robert Mitchell, ingenious person has repaired it ("Puggy") so beautifully and correctly. No attempt has been made to give ornaments to the stone which it formally wanted. Mr Mitchell has carefully attended to the original carving, compared it to wear worn by time with the representations of the Scottish Arms in books of heraldry and added nothing to the stone itself."* This surely constitutes today, sound conservation practice. Indeed it is evident that Mr Mitchell undertook careful and sensitive paint analysis.

In 1995, the Mercat Cross was given a thorough photographic record *in situ* prior to any works. Once back in the studio at Edinburgh it became evident that a number of decay mechanisms existed within the subject. The colonisation by organic growth for example which by its existence produces oxalic acid and is hygroscopic holding moisture within the pore structure of the stone, making the substrate susceptible to freeze thaw action. Also evident are the effects of copper sulphate wash as a result of rain water running from the cuprous horn upon the unicorn's head, which had served as an effective biocide but also introduced salts to the forehead of the subject as seen. The Royal Commission in Scotland describes *"Puggy"* as posed in a sitting position and wearing a collar from which a chain is carried around the body. In front of the breast there is a crowned shield bearing the Royal Arms of Scotland, and this is surrounded by the collar of the Order of the Thistle, the figure of St Andrew with the cross can be seen on the badge. It is interesting to note that repairs apparently conducted by Mr Mitchell in 1830 upon the collar and badge of St Andrew still remain 150 years later. What is more surprising is that upon examination, those repairs were found to have been modelled in lime. Further polychrome in the form of red and white pigment was still visible upon the rampant lion on the *"Puggy's"* crowned shield. Pictured within Figure 2 the unfortunate column which was broken by contractors who employed an inappropriate removal method namely the backhow of a JCB. The column was repaired by the Centre and re-erected in Stirling to accommodate *"Puggy"* and complete the conservation of the Stirling Mercat Cross.

3. Case Study: Rosslyn Chapel

A study of the Centre's records on Rosslyn Chapel indicated a fascinating history of conservation works and provided sufficient documentary evidence to allow an assessment of past and current conservation practice.

The north clearstory cornice of St Matthew's Episcopal Chapel in Rosslyn features the date 1450. It is considered that this represents the date that Rosslyn

Chapel was established by William Sinclair the Third Earl of Orkney. Documentary evidence regarding the building of the Chapel exists in *'Scotichronicon',* a 1447 journal, wherein it is mentioned *"Lord William Sinclair is erecting an elegant structure at Rosslyn".* The Chapel is, however, incomplete. Of the cruciform design only the choir and parts of the east walls and transepts were built. It is of course speculation that the incomplete nature of the Chapel may be due at least in part to the highly decorative and elaborate carved stone work throughout the building and the associated costs therein. The carved interior stonework is one of the main features of Rosslyn Chapel. Christopher Wilson states, within the Lothian Edition of the Buildings in Scotland series, "*the sculpture has a denseness and repititiousness that resembles cake icing or topiary more than carving in stone".*

The choir roof here is of a pointed tunnel design and rises to 12½ metres from floor to apex. The roof is divided into five bays, each separated by ribs decorated with pendant cusps. The cross ribs of the roof have considerable projection with *Fleurs-de-lys* and other vegetal forms. The panels which are divided by the ribs are themselves richly decorated with each panel featuring a single motif in bold relief, for example as we see here, one panel may feature star images while another feature roses, another squares and so on. Wilson described construction elsewhere within the Chapel as having an overwhelming visual impact. As described, the ribs feature heavy cusps which hang from strips of foliate moulding. The transverse ribs feature two orders giving an extraordinary thickness. Wilson describes the pendant bosses of the intersections as having a disagreeable sagging profile. Even more disturbing are the fat pendants of the springings that point diagonally downwards. The ambulatory and the Lady Chapel are divided by elaborate piers known as the Mason's or Earl's Pillar and the Apprentice Pillar. The Apprentice Pillar is illustrated within Figure 3.

The Apprentice Pillar - the South Pier is particularly notable for as the legend goes

"the Master Mason having received from his patron the model of a pillar of exquisite workmanship and design hesitated to carry it out until he had been to Rome or some such foreign part and seen the original. He went abroad and in his absence an apprentice having dreamed of the finished pillar at once set to work and carried the design as it now stands. A perfect marvel of workmanship. The Master Mason on his return was so stung with envy that he asked who had dared to do it in his absence on being told that it was his own apprentice he was so inflamed with rage and passion that he struck him with his mallet, killed him on the spot and paid the penalty for his rash and cruel act".

Figure 3. The Apprentice Pillar in Rosslyn Chapel.

4. Discussion

Having given a description of the highly decorated interior stone work it would be appropriate to examine previous works undertaken now to repair and conserve. It is possible only to make a passing reference to works undertaken during 1870-71 by David Bryce due simply to the lack of documentary evidence. McWilliam, within *The Buildings of Scotland*, states that accounts of these works were given in two contemporary publications - *The Builder* and *The Building News*. *The Builder*, he states, reports that all badly decayed stones were replaced and almost the whole of the carving was retooled and sharpened. McWilliam continues by stating that *The*

Building News gave a more emotive account: parts of the stone cleaned with acids, others rechiseled, destroying the original proportions of the mouldings and altering entirely the character of the ornament. A third of the cusps on the nave vault were restored in cement.

It is appropriate to consider the work undertaken upon the interior stone work by the Ministry of Works and Ancient Monuments Branch, Historic Scotland's predecessor between the years 1954 and 1957. Lord Rosslyn was the client. During the months of February and March 1954 the Ancient Monuments Division were employed to clean and preserve the interior stone work of the Lady Chapel. This work was undertaken on a recoverable basis, that is the sum of £110 was charged to the client for the work. At the end of March 1954 as the first phase of work was nearing completion, Lord Rosslyn wrote via his Agent, Captain MacIntyre, to express how pleased he was with the work carried out by the Ministry at Rosslyn Chapel and hoped *"that they may be allowed to continue the good work next winter" on the basis of this request the Ancient Monuments Division produced a report which set out and gave costings for further works throughout the rest of the interior of the Chapel. The report produced in May 1954 is titled "Proposed Further Internal Cleaning and Preservation of Stone work and Ornament".*

The report interestingly was written in a format which is still in use today giving a description of the subject providing an account of the condition of the subject and finally proposing and recommending the appropriate conservation process. In these terms the report is thoroughly modern.

Within the recommendations it was stated *"that surfaces covered with green algae will be scrubbed down with stiff bristle brushes using a solution of ammonia and water and the operation repeated until all evidence of coloration disappears. Water will then be used copiously until the surfaces are clean and free from dirt and vegetation. Great care will be exercised during the scrubbing process to avoid damage to fragile areas. Treated surfaces will be allowed to dry and examined for hollow or blown areas where no ornament occurs flaky patches will be scaled off. Hollow areas on ornament will receive special treatment by grouting and pinpointing to consolidate the subject and preserve the design outline. When the surfaces are thoroughly dry they will be hardened with silica fluoride of magnesium at a rate of 1lb per 2 gallons of water (approximately 1kg per 20 litres), three or four coats will be necessary but the treatment will be continued until hardness is achieved. Tinting of the stone will then follow and finally a waterproofing treatment of shellac and methylated spirits."*

It has already been mentioned that the form of the 1954 report complies with much of today's conservation philosophy. It is pertinent to consider how appropriate many of the recommendations contained within this report are in terms of current conservation philosophy. The report recommends a full photographic survey which was conducted prior to any work being carried out. A photographic survey was also undertaken of the completed work. This together with the wealth of written documentation left by the Ancient Monument Branch demonstrates the importance given to record keeping at the time. Today, of course, record keeping, both photographic and written, for archival purposes is fundamental to any conservation process. As a practising conservator it is extremely helpful to approach any work with a full knowledge of those processes undertaken in the past. So often, however, the conservator is left to proceed without full knowledge and as a result must expect the occasional nasty surprise.

The use of an ammonia solution for cleaning is not a process that would be employed today, if for no other reason on the grounds of health and safety. Conservators will also be reluctant now to use the large quantities of water used in the 1950s on a church interior, that is not to say however, that I can criticise the use of these methods in the context of the 1950s when conservation was, perhaps, in its infancy. The fact that these processes were employed and documented has actively contributed to conservation processes which are deemed acceptable today.

The removal of cement repairs and pointing by the Ancient Monuments Branch and subsequent repair with what is described as a synthetic stone is wholly acceptable by today's standards. Within the report, reference is made to the repair of small areas with a synthetic stone of matching colour and texture. It is evident that the Ancient Monuments Branch were manufacturing a repair material. The design and manufacture of a repair medium, whether it be lime based or Paraloid B-72 based must be sympathetic to the surrounding stone in terms of texture, colour and durability, this is a fundamental premise within the field of stone conservation today.

The application of silica fluoride magnesium; a cementitious slurry can be viewed as a use of new technology of the time. In 1995 it can be seen that the application of such a material is to the detriment of the interior stonework as soluble pollutants are concentrated in the stones substrate and are unable to disperse due to the impermeable nature of the applied material. Having identified the material by the documentary evidence the results of removing this material are far easier to predict. Indeed the records show that the conservator can expect to rediscover painted decoration in the Choir roof as this was recorded before being obscured by the surface treatment. The use of shellac in a solution of methylated spirits by the Ancient Monuments Branch in 1954 can perhaps be seen as a precursor to the modern conservator's use of acrylic resin such as Paraloid B-72 in a solution of acetone. A significant attribute of both is their reversibility. Having identified the material within Rosslyn Chapel it is a relatively more simple exercise to remove it.

In 1995 the interior stone work of Rosslyn Chapel was once again in poor condition. The fabric of the building was under attack from descending moisture and associated problems thereof. Heavy recolonisation by organic growth was evident within the choir roof and along much of the north aisle. An environment data logger located within the chapel since 1995 has consistently recorded a relative humidity of over 80%. The descending moisture and high humidity within the chapel are being addressed presently as a temporary roof has been constructed over the chapel in order to dry the fabric of the building. There are of course potential dangers inherent in drying out a building which has for so long been subject to moisture. In February 1956 the Ancient Monument Branch sent a sample of efflorescence to a consultant scientist, who identified the efflorescence as sodium bicarbonate. An example of 1990's efflorescence was found breaching the hardened surface. With an impermeable barrier effectively covering the entire surface of the interior stonework the danger of cryptofluorescence and consequential loss of carved detail is a real one. During the early part of 1995 a demonstration of cleaning by Nd:Yag Laser equipment was undertaken to remove numerous layers of surface treatment applied in the 50's. The laser treatment was extremely effective in removing the material and was non-invasive.

5. Conclusion

During my involvement as Stone Conservator of the Rosslyn Chapel and the associated research into those works undertaken by the Ministry of Works, I have been made aware that conservation, specifically stone conservation, is not a recent phenomenon. It is possible to trace specific methods which conservators employ today directly back to methods undertaken 50 years ago. It has been the diligent record keeping of past conservators which has allowed this lineage to be identified. Consequently, the present generation of conservators has a responsibility to record their works so that future conservators may trace that lineage further still.

Stone Weathering and Atmospheric Pollution Network'97: Aspects of Stone Weathering, Decay and Conservation.
Edited by M.S. Jones & R.D. Wakefield © 1998 Imperial College Press.

THE ROLE OF TECHNICAL CONSERVATION RESEARCH AND EDUCATION DIVISION HISTORIC SCOTLAND

N.M. ROSS
Technical Conservation, Research and Education (TCRE)
Historic Scotland, Edinburgh, U.K.
Tel 0131 668 8600. Fax 0131 668 2669

As the country's leading conservation body, Historic Scotland's aims are to protect Scotland's built heritage and to present this heritage to the public, to advise the Secretary of State for Scotland on policy matters and to manage properties in his care. Additionally, Historic Scotland fulfils statutory obligations to protect ancient monuments and historic buildings in the ownership of others. Through an integrated approach Historic Scotland aims to encourage industry and the education and training providers to achieve this objective. Against this framework Historic Scotland established a Technical Conservation, Research and Education Division (TCRE) in 1993. This paper introduces the objectives of TCRE and describes current research projects funded by Historic Scotland.

1. Introduction

Historic Scotland maintains, conserves and presents some 330 properties in care, and has a commitment to educate and disseminate knowledge about the built heritage and its conservation. Under legislation, the Secretary of State has statutory powers and duties which are administered by Historic Scotland to schedule ancient monuments and to list buildings of special architectural or historic interest. Some 6,000 scheduled monuments and 40,000 listed buildings are on the current lists.

Within the statutory framework, protection for the built heritage is secured through consent systems for scheduled monuments, listed buildings and buildings in conservation areas. To support this, financial incentives can be offered to owners in the form of grants for the repair of historic buildings and the protection of monuments. Currently this amounts to £12 million per annum. In providing financial assistance for repairs to around 140 outstanding buildings in private ownership each year, and through pragmatic conservation work on the 330 monuments in State guardianship, Historic Scotland has gained wide experience of the need for, and use of, high quality traditional building materials and associated craft skills.

2. The role of the Technical Conservation Research and Education Division (TCRE).

To undertake effective conservation work it is crucial that properly trained craftsmen have appropriate traditional building materials at their disposal. Through an integrated approach Historic Scotland aims to encourage industry and the education and training providers to achieve this objective. Against this framework Historic Scotland established a Technical Conservation, Research and Education Division (TCRE) in 1993. The Division is charged with the responsibility for researching appropriate issues, developing skills relating to the built heritage, and raising the standard of conservation practice among owners, contractors and professional groups.

To achieve these aims there is a need to:-

- Undertake a programme of research relating to historic structures in conjunction with others where appropriate.
- Disseminate the outcome of scientific conservation research through a structured and integrated approach to publications, lecturing and training.
- Develop and improve training in conservation skills with Scottish educational institutions and colleges.
- Devise standards for accreditation of professional competence and quality control.
- Investigate the supply of traditional building materials to encourage the revival of key industries.

To be successful in the future, conservation projects must be driven by an integrated and research-based understanding, improved education and training, and the assurance that appropriate materials are going to be available when required. The Division is structured to carry out these tasks in three inter-related branches.

2.1 Branch 1

Branch 1 of TCRE is based at the Historic Scotland headquarters at Longmore House, Edinburgh and provides scientific conservation, research, analysis and dissemination, technical advice, education and training. The ongoing programme of technical publications and research projects commissioned from universities and research organisations includes the following:-

2.1.1 Timber decay

An investigation is underway at the Scottish Institute for Wood Technology at the University of Abertay, Dundee, into issues associated with dry rot in historic timbers and its control and treatment by environmental methods. The research will examine the effects of air flow, moisture content and temperature on the growth and decay of dry rot and its organism *Serpula lacrymans*. A research report, PhD thesis and Technical Advice Note (TAN) will result from this project.

2.1.2 Scottish limestone

Research has been commissioned to evaluate a range of selected Scottish limestones, with a view to their preparation and use as building limes. The outcome of this work which includes fieldwork, sampling, testing and analysis, is expected to be presented in a final report by the end of 1997.

2.1.3 Historic plasterwork

Historic Scotland has undertaken a project to reinstate the Royal Apartments at Edinburgh Castle. The project includes an investigation of 17th Century plasterwork and plastering techniques, the preparation of moulds and sample mixes and establishing original methods of work and plaster specifications. This research will form the basis of a Technical Advice Note on historic lime plasterwork. The project provides for the training of two local plasterers in historic plastering techniques in order that the acquired knowledge and skills can be passed to others in the future. The research, training period and contract works will be filmed and photographed and will be used in the preparation of a training video for conservation professionals of all disciplines.

2.1.4 Stonecleaning

Following from an earlier research project by The Robert Gordon University (RGU) Aberdeen into cleaning of granite buildings [1], a Technical Advice Note on the Stonecleaning of Granite Buildings is being prepared for publication in mid-1997. In a similar vein a joint Building Research Establishment (BRE) and RGU research team (with input by Historic Scotland's Stenhouse Conservation Centre) is investigating issues associated with laser cleaning of sandstone. This includes cleaning trials; the economics of laser cleaning; health and safety aspects and the development of scanning systems. Recent examples of laser stonecleaning have been examined and methods and results assessed. The research findings will be submitted to Historic Scotland as a final report in 1998.

2.1.5 Sandstone consolidants and water repellents

A joint BRE and RGU research team is taking forward research begun in the 1970s and early 1980s into the application of masonry consolidants to monuments and historic buildings to arrest surface decay from natural weathering. The programme includes the testing of currently-available proprietory treatments on masonry panels. Cores have been taken for laboratory analysis of consolidant and waterproofer absorption, colour change and strength tests. Monument test sites previously treated with "Brethane" have been revisited for further inspection, and Stenhouse Conservation Centre has contributed specialist guidance on the application of treatments. The programme is due for completion in 1998.

2.1.6 Traditional materials and building techniques

A greater understanding of the wide variety of natural materials used in traditional Scottish thatching has steadily developed in recent years. Much of this knowledge has been rediscovered as a result of grant-aiding thatched buildings around the country under Historic Scotland's Thatched Houses Maintenance Scheme. This has resulted in better understanding of traditional methods and materials and as a result, a Technical Advice Note (TAN) titled Thatch and Thatching Techniques (TAN4*)* was published [2] followed by The Hebridean Blackhouse (TAN5) [3] in the same year. This included a detailed study of the house at 42 Arnol, Isle of Lewis, which is in state care (Figure 1).

The blackhouse had recently undergone major conservation work and the opportunity was taken to write up the project for the benefit of others proposing to carry out repairs to these buildings. Earth Structures and Construction in Scotland (TAN6) [4] gathered together the collective knowledge of rural structures and materials. The use of earth in Scottish building construction ranges from the earliest prehistoric times to the 1950s and a number of common denominators are known to exist. Having amassed a considerable volume of knowledge from a number of hands-on conservation projects, it is recognised that much experimental work still remains to be carried out under scientific scrutiny.

Figure 1. Blackhouse at Arnol, Lewis, following reconstruction. A building type which evolved over hundreds of years utilising indigenous, and renewable materials.

2.1.7 Lime mortar

A project commenced in August 1995 by the University of Paisley aims to develop a methodology for researching the micro-interaction of traditional lime mortar and stone. It is intended that further research will provide guidance on the selection of lime mortars, and on their production and application for conservation professionals and craftsmen.

2.1.8 Slate use and performance

Research underway at the University of Dundee is investigating the supply and use of traditional Scottish slate in the repair and restoration of buildings. The programme, due for completion in late 1997, includes comprehensive surveys of buildings in selected conservation areas. The ensuing TAN will provide much-needed guidance on the use of Scottish slate for the benefit of practitioners, roofing contractors and owners. A complementary project is underway at Glasgow University to investigate the performance of Scottish slate and to assess the extent of slate resources. This aims to identify the distinctive qualities of Scottish slate and to quantify the reserves with a view to encouraging a revival of the industry. The study will update the last comprehensive survey of the Scottish industry carried out in 1944 [5]. The project will be completed in late 1998.

2.1.9 Carved stone decay

The Building Research Establishment is devising a research-based methodology for recording the decay of carved stones throughout Scotland. It is anticipated that a separate Stage 2 project may start during 1997, undertaking a selective survey of Scotland's carved stones (Figure 2) using the established methodology. Stage 3 would then consider what range of conservation solutions might be relevant to the identified and quantified problems.

Figure 2. The Maiden Stone, Aberdeenshire, a red granite, upright cross slab bearing Pictish, Christian and decorative symbols, much of which has been lost by weathering.

2.1.10 Graffiti removal

Research being carried out by BRE in partnership with the Stenhouse Conservation Centre is assessing the effects of physical and chemical cleaning techniques for the removal of graffiti from different types of stone. The research findings will be published as a TAN.

2.1.11 Oil pollution

Following the Braer tanker shipwreck in Shetland in 1993 a study by The Robert Gordon University, Aberdeen, is assessing the effect of oil pollution on the Historic Scotland site of Jarlshof over a 5 year period. The study is due to be completed in November 1998.

2.1.12 Masonry biocides

Building upon original research work carried out for TCRE, The Robert Gordon University study into biological growths and the application of biocides to sandstone buildings in Scotland [6] will be published as a Technical Advice Note in early 1998. This will offer guidance to practitioners and others on the control and treatment of surface growth on masonry.

2.1.13 The Duff House project report

Following completion of the Duff House Country House Gallery Project, Banff, in 1995, there was a need to report on the repair and conversion of this William Adam mansion for the benefit of others undertaking major projects of this kind. (Study due for completion, end of 1997). In addition to the report Historic Scotland aims to produce a series of training videos on the range of conservation techniques employed during the works with particular emphasis on fire protection measures.

2.2 Branch 2

The Historic Scotland Conservation Centre, based at Stenhouse Mansion and South Gyle, Edinburgh, constitutes Branch 2 of TCRE and carries out specialised practical conservation work on easel paintings, structural paintings and masonry conservation. The Centre aims to provide relevant pragmatic advice to building professionals and owners. The programme of work for Properties in Care Division of Historic Scotland has included conservation of the Stone of Destiny prior to its installation at Edinburgh Castle, conservation work at Holyrood Palace, Stirling Castle, Elgin Cathedral, the Bruce's Heart casket at Melrose Abbey, and Sueno's Stone. The Centre also provides advisory reports to building owners on a wide variety of building and artefact conservation problems.

2.3 Branch 3

The *Scottish Conservation Bureau*, based at Longmore House, Edinburgh, alongside Branch 1 provides a conservation information database and grants resource centre for Scotland. It funds internships for conservators and offers assistance in the development of their skills and experience.

3. TCRE Information Dissemination Framework

TCRE's intention is to continue to work to a pre-planned strategy where published Research Reports, Technical Advice Notes and Practitioners Guides will be the products of a wide-ranging research programme appropriate to the needs of conservation professionals. All the research findings will be made available to interested parties. These include:

- Building professionals and agents and the Building industry
- Education and training establishments
- Technical libraries
- Conservation bodies
- Official bodies
- Others working in the field
- Building owners
- Historic Scotland staff and associated groups
- International Conservation links

4. Conclusions

In taking a key role in the promotion of the use of appropriate traditional materials, Historic Scotland is playing a major part through its Historic Buildings Repair Grant Scheme to pump-prime the industry in this area. With repair and maintenance in Scotland now amounting to over 50% of all construction spending, this input has considerable influence. Through working in partnership with others, Historic Scotland believes that a greater and better understanding of the issues will result. However, no matter what the approach, success will be judged by the effectiveness of the practical decisions taken in actual work on the historic building stock.

By promoting the results of its research programmes through publications, conferences, seminars and lectures, Historic Scotland also aims to take forward with industry, the professions, and the education and training providers an integrated approach to the future development and use of traditional Scottish building materials. In the anticipated success of so doing, the value of Scotland's protected historic building stock should remain high.

5. References

1. Masonry Conservation Research Group, The Robert Gordon University. *Report of the Research Commission investigating Cleaning of Granite Buildings*, (Historic Scotland, Edinburgh, 1995).
2. Walker, McGregor and Stark., *Thatches and Thatching Techniques, A Guide to Conserving Scottish Thatching Traditions*, (Historic Scotland, Edinburgh, 1996).
3. Walker and McGregor., *The Hebridean Blackhouse, A Guide to Materials, Construction and Maintenance*, (Historic Scotland, Edinburgh, 1996).
4. Walker, McGregor and Little., *Earth Structures and Construction in Scotland, A Guide to the Recognition and Conservation of Earth Technology in Scottish Buildings,* (Historic Scotland, Edinburgh, 1996).
5. Richey J.E. and Anderson J.G.C., Scottish Slates, *Wartime Pamphlet* No **40**, (Department of Scientific and Industrial Research, 1944).
6. Masonry Conservation Research Group, The Robert Gordon University. *Report of the Research Commission on biological growths and the application of biocides to sandstone buildings in Scotland* (Historic Scotland, Edinburgh, 1994).

Stone Weathering and Atmospheric Pollution Network '97: Aspects of Stone Weathering, Decay and Conservation.
Edited by M.S. Jones & R.D. Wakefield © 1998 Imperial College Press.

STONEWORK CONSERVATION IN THE NATIONAL TRUST FOR SCOTLAND

I.M. DAVIDSON
Regional Buildings Surveyor (Grampian), The National Trust for Scotland
Grampian Regional Office, The stables, Castle Fraser, Sauchen
Inverurie, Aberdeenshire, AB51 7LD
Tel 01330 833225 Fax 01330 833666

In the North East of Scotland the principal building stone is granite which appears in a number of different forms, both in terms of construction and colour. Granite is a dense material which suffers rarely from atmospheric pollution in this part of Scotland. Where deterioration is found it is usually in the form of sugaring on the pinker stones as a result of mechanical damage such as foot traffic. However, there are a number of sandstone buildings towards the North of this area particularly around Turriff and Elgin where both pink and yellow sandstones have been quarried and used. Furthermore, decorative panels, projections and finials are found on all buildings in sandstone. Limestone is almost unknown in the area as a building stone, although there is a substantial band of this material running through the region from around Portsoy through to Keith. In recent years there has been some research into the history and use of this limestone. It is now clear that there were in excess of 40 quarries operating at the peak of the extraction ranging from small pits to very large extractions. However, most of this material was being used for agricultural purposes and following improvements in transport infrastructure better quality lime from elsewhere, particularly England, became available and was preferred to the relatively inaccessible and hard won North-east limestone. These remote quarries subsequently fell into disuse and were closed. There are now only three in operation and all produce lime for agricultural purposes only. However, it would appear that the clayey limestone available from these sources would have produced a good building lime and there are small scale experiments being undertaken now to determine the usefulness of this material for conservation work.

Stonework conservation is firstly, the conservation of individual stones through application of consolidants or protective covers. Secondly, the conservation of built structures in either complete or ruined form. In either case there must be a philosophical dimension to any intervention attempting to prolong the life of the stonework in question. There is no shortage of printed material available now for those involved in such conservation work. Technical issues are well covered by such bodies as English Heritage and Historic Scotland. It is now usual to find publications such as the Practical Building Conservation handbooks published by English Heritage and the Historic Scotland Technical Advice Notes on the bookshelves of caring practitioners. However without a sound understanding of the philosophical issues associated with the use of modern techniques in both stonework conservation and other forms of building conservation there is a danger of misapplication and a subsequent diminution of the built heritage in both the macro and micro environment.

Over recent years the National Trust for Scotland has wrestled with these philosophical issues and it is clear that it has not operated in isolation. A number of other bodies have faced these issues squarely and well written and considered documents have been produced. Perhaps foremost among these are those developed by ICOMOS in its various forms. These have followed on from and developed the clearly understood philosophies underpinning the Society for the Protection of Ancient Buildings both as described by William Morris in his manifesto and subsequently developed and enhanced by A R Powys in his publication "Repair of Ancient Buildings", recently republished. Throughout all of these documents there is an important consideration for the conservation of stonework which is that the existing fabric should not be disturbed wherever possible and that the patina developed through time and use should be left in place. The visible development of the history of the stonework should not be compromised by any conservation work and that only materials that are sympathetic to the original fabric should be used. That would not exclude the use of contemporary materials but they should be proven to be compatible with the original.

In all cases decisions on conservation work must not be confined to a thorough understanding of the existing fabric (the nature of the material itself and the causes of its decay). The history, development and previous use may also significantly affect the decision making process, determining the nature of conservation work, or even whether it is desirable at all. Perhaps the most contentious issue that stonework conservators might be faced with is the conservation of ruined masonry structures when a small cottage or large cathedral has lost its protective wallhead coverings and has perhaps been subjected to the ravages of "quarrying" by local people. The destructive effects of wind, rain and frost on the degraded structure can become dangerous and its place in the landscape can be put at risk resulting in a demand for conservation work with ensuing technical and philosophical complexities.

There is a long history of conservation work on important ruined structures and these can be seen throughout the land. Techniques associated with the careful conservation of these ruined structures are well established and visible in many properties cared for by Government Agencies. In addition caring private owners undertake similar works often with the advice of these agencies. Perhaps the philosophical issues associated with such work are now taken for granted. However, it might be argued in many instances this work has damaged the aspects of the place which were once found important. Ruined structures are often considered important not only for their historic value but also for their cultural and aesthetic qualities. A crumbling and decrepit ruin can have an evocative picturesque contribution to the cultural landscape with its broken outline echoing past traumas and gentile decay. Conservation work, which essentially halts this process is perhaps contributing to falsification, limiting the importance in the landscape of these places.

It would be hard now to argue against such conservation work on structures such as Fountains Abbey or Elgin Cathedral. However, where the ruined structure is a Highland Settlement (cleared in the late 18th/early 19th Century) this becomes an important part of our understanding of the social and economic changes in the use of land and the population that once occupied it. The broken appearance of this unconserved structure with its patina of age is shown by the vegetation growing over it, and through it. Thus a hard cleaning and consolidating procedure that holds the ruined building in place for all time in such a situation could well be an inappropriate intervention that diminishes the significance of these places. These factors should be well considered by the Conservator prior to any intervention work.

Other important technical issues are becoming apparent where careful consideration of the philosophical issues should also be considered. For example, where the original construction of stonework introduced a decay mechanism that will ultimately destroy the material, such as the use of ferrous cramps to hold together stone in a decorative armorial from the early 20th Century. Perhaps part of the authentic fabric should be removed to save the significant part from destruction. A strong argument can be constructed for such work and perhaps by adequately and accurately recording both the existing construction together with the conservation work this intervention can be considered acceptable. Article 20 of the Burra Charter perhaps assists in understanding how such an intervention can be justified. It states "adaption (for example the removal of iron cramps) is acceptable where the conservation of the place cannot otherwise be achieved, and where the adaption does not significantly detract from its cultural significance". The Charter goes on to expand and explain the terms 'adaption' and 'cultural significance' and an important part of determining the cultural significance is through the creation of a conservation policy. The Conservation Policy can be prepared for either a small piece of repair work or a major building or town scheme. The main criteria contained within it would follow similar guidelines. The policy would assess all the information relevant for future conservation and presented in three clear sections:

a. Information on the history and development of the place

b. A statement on the importance or cultural significance of the place

c. A statement on how this significance would be conserved

Using such a technique a Conservator would be able to construct a defensible solution to the conservation problem.

Stone Weathering and Atmospheric Pollution Network '97: Aspects of Stone Weathering, Decay and Conservation.
Edited by M.S. Jones & R.D. Wakefield © 1998 Imperial College Press.

CARVED STONES

IAN A.G. SHEPHERD
Aberdeenshire Archaeologist
Aberdeenshire Council, Woodhill House, Aberdeen.

The protection of carved stones remains, on first sight, deceptively simple. The stones, of a variety of dates, suffer from varying rates of attrition according to the nature of their parent material, location, past conservation history, current management, etc. Some of the stones are symbols of the merging of two cultures, the Picts and the Scots, in the 9th century. It is true that progress has been made in recent years in developing an understanding of the forces acting upon these stones, and in spreading a cautious, admonitory message regarding attempts at conservation. In this regard the work of the National Committee on Carved Stones in Scotland in publishing (in conjunction with Historic Scotland) 'Photographing Carved Stones: a practical guide to recording Scotland's past' [1] is particularly apposite. Here will be found not only precise advice on angled flash photography from the master photographer himself, but also (justified) warnings against any 'contact' operations such as lifting or lichen cleaning by inexpert hands. However, successful examples of the application of the (long-standing but still active) government policy to move stones under cover are few and far between. This comparative lack of success may stem partly from a confusion of roles that these stones are asked to perform: local heritage attractions, art objects, national symbols, landscape features, institutional art - all can be illustrated by current cases in Scotland. The extraordinarily protracted saga of the Dupplin Cross has finally been settled by an uneasy compromise; the Aberlemno stones remain in need of shelter; the Dunnichen stone was 'repatriated' to Angus from its safe-keeping in Dundee Museum; the Maiden Stone awaits a cover. The conservation of such material requires a balance between local needs and the interests of the stone. In all these cases, the views of the local communities, powerfully if sometimes contradictorily expressed, have been an important factor. Looking to the future, the precautionary principle is being applied, rightly, with greater consistency. Rubbing, chalking, wetting are all discouraged, although further, non interventionist, recording is encouraged. The Treasure Trove system is developing a role in the protection of newly discovered stones. The opportunities for new local centres or 'lapidaria' are being re-examined, while selected monuments will be, albeit painfully slowly, afforded greater shelter.

References

1. Gray T. and Ferguson L., *Photographing Carved Stones: a practical guide to recording Scotland's past'* (Pinkfoot Press, 1997)

Stone Weathering and Atmospheric Pollution Network '97: Aspects of Stone Weathering, Decay and Conservation.
Edited by M.S. Jones & R.D. Wakefield © 1998 Imperial College Press.

EFFECT OF HYDROCARBONS ON BIOFILM DEVELOPMENT ON SANDSTONE

R. V. YORDANOV, K. NICHOLSON
Environmental Geochemistry Research Group,
The Robert Gordon University,
St. Andrew Street, Aberdeen.

Hydrocarbon contamination of the environment has been a widespread phenomenon during the last decades with several severe oil-spill disasters. Oil is also continuously released through the bilge and fuel oil discharges of vessels. This type of contamination accounts for almost 0.3 million tons of the annual hydrocarbon input to the worlds oceans [1] whilst the overall annual global input of hydrocarbons (HC) is estimated to be approximately between 1.7 and 8.8 million metric tonnes. Some of the oil can reach buildings in the viscinity of the discharge, as occurred in Shetland with the *Braer* spill.

Biological growth on stone is already a recognised factor in building deterioration and is of a particular concern with monuments and cultural heritage sites. Jones *et al.* [2] have evidence showing that both physical and biochemical mechanisms are involved in stone deterioration. Wakefield [3] also observed that some organic by-products of the microbial activity can chelate minerals from the sand stone thus changing the stone properties. Other authors [4, 5] have indicated that the HC can be used as carbon source by microrganisms and this ability has been extensively utilised for conducting bioremediation experiments. The main objective of the present work was to assess the possible effect of hydrocarbons on the biofilm growth on sandstone and its potential deleterious effect on the buildings.

In a series of laboratory trials, the effect of HC on microbial growth was studied. The methods selected as suitable for the purpose of this study were used for quantification of biofilms growing on the sand particles and were described elsewhere [6]. The changes in autotrophic biomass are being measured by using hot ethanol extraction of chlorophyll-a, and subsequent determination at 665nm. The use of laser-induced-fluorescence method is also under consideration. The number of viable bacteria is being determined as colony forming units (CFUs) on R2A growth media. Measurement of the overall biomass is being carried out using a modified method of Lowry *et al.* [6] cited in Yordanov *et al.* [6] for cell protein determination.

The chlorophyll levels were still very low and no differences were observed during the first 4 months of the biofilm development.

Epilythic microbial biomass was measured during the early colonisation and growth phase of its development. The initial results showed higher bacterial and overall biomass in oil enriched environment. The CFUs increase almost 20-fold since the beginning of the experiment in the HC samples (8.3×10^7) and were twice as high than the control group. The same trend was observed for the cell protein measured as Folin reactive material (FRM) which was $2.4 mg/cm^{-2}$ stone surface in HC samples, and $1.2 mg/cm^{-2}$ in controls. This is probably a reflection of the increased amount of available carbon on the sandstone surface. The availability of HC to the micro--organisms depends on many factors such as HC concentration, ambient temperature, oxygen, nutrients etc. The degradation of HC is also result of their composition, and generally saturates are characterised by the highest degradation rate, although other options are not excluded. The increased biofilm growth may affect the sandstone either directly by release of metabolites, or by changing the physical structure of the stone during fluctuations in the ambient temperature (i.e. freeze\thaw conditions) which may be linked to the production of extracellular polymeric substances. Therefore higher biofilm growth may be expected to deteriorate to a higher extent on the sandstone.

References

1. National Resource Council. *Oil in the Sea: Inputs, Fates, and Effects*, (National Academy Press, Washington, D.C. 1985)
2. Jones M. S. Wakefield R.D. Forsyth G., *Biodeterioration of stone by algae and application of stone conservation methods.* EPSRC Report GR/J91500, (1996).
3. Wakefield R.D., Personal Communication, OptoElectronics Research Group, The Robert Gordon University, Aberdeen.
4. Owen D., Bioremediation of marine oil spills: Scientific validity and operational constraints. Arctic and Marine Oil Spill Program Technical Seminar, Vancouver, BC, Canada, (June 1991), 119-130.
5. Von Wedel R.J. *et al.* Bacterial biodegradation of petroleum hydrocarbons in groundwater: in situ augmented bioreclamation with enrichment isolates in California. *Wat. Sci. Tech.*, **20**, 11/12, (1989) 501-503.
6. Yordanov R.V *et al.* Biomass characteristics of slow sand filters receiving ozonated water. In Advances in Slow Sand and Alternative Biological Filtration ed. Graham, N.J.D. and Collins M.R,. (John Wiley & Sons, Chichester, 1996) 107:118.

Stone Weathering and Atmospheric Pollution Network'97: Aspects of Stone Weathering, Decay and Conservation.
Edited by M.S. Jones & R.D. Wakefield © 1998 Imperial College Press.

HEAT AND MASS TRANSFER AT THE SURFACE OF TWO SANDSTONE TYPES UNDER DIFFERENT ATMOSPHERIC CONDITIONS

S. POMBO FERNANDEZ AND P. MARTIN
Masonry Conservation Research Group
The Robert Gordon University, Aberdeen, AB25 1HG, Scotland.
Tel ++(0)1224-262000 Fax ++(0)1224-162828
E mail s.pombo@rgu.ac.uk

Daily temperature-humidity cycles were produced in a climatic chamber which contained samples of two sandstone types identified as Crossland Hill ('blonde' sandstone), a quartz arenite; (9.7% porosity), and Locharbriggs, ('red' sandstone), a ferruginous quartz arenite; (24% porosity). Samples were heated and cooled at regular periods in the chamber. The difference between stone surface temperature and the ambient air temperature was monitored by using thermo-photometry, which makes use of the infra-red spectral band. The infra-red scanner converts electromagnetic thermal energy radiated from an object into electronic video signals. Results showed that although surfaces of the sandstone samples had the same size, shape and relative orientation, they did not emit or absorb the same amount of thermal radiant energy. This could be explained by differences in the surface roughness and colour measured by colour meter (Red sandstone, a*=7.4; Blonde sandstone, a*=1.7). Here, the temperature distribution of the two sandstones at peak periods of each cycle sharply contrasted. The exposed test stone surface was colder than the surrounding air, on average, during the 35°C, 70% cycle, for both sandstones.

Background

Measurements of temperature and humidity at, above or within stone surfaces is very difficult to achieve in real time without the measuring instruments used unduly influencing the results obtained. This work illustrates the results of a series of surface temperature measurements taken by Infra-Red Thermography, which is able to demonstrate changes to resultant stone surface temperatures when exposed to cycles of ambient temperature and humidity executed under controlled laboratory conditions. Using this technique it is possible to distinguish between the behaviour of different stones under similar conditions and to speculate as to the surface processes that bring about variations in the surface temperatures observed. Although the camera is only able to detect temperatures at a surface, it does so at a high resolution in real time and should be able to detect the presence of convective cells, which are a possible mechanism under the forced convection regime used in the tests. The application of this technique to stone in the field may prove more difficult, but the observations made with this novel experiment have shown to be consistent and reproducible.

Stone Weathering and Atmospheric Pollution Network'97: Aspects of Stone Weathering, Decay and Conservation.
Edited by M.S. Jones & R.D. Wakefield © 1998 Imperial College Press.

LASER CLEANING OF SCOTTISH SANDSTONE AND GRANITE

R.D. WAKEFIELD, D. MCSTAY, E. BRECHET,
The Optoelectronics Research Group
The Robert Gordon University
Aberdeen. AB25 1HG. Scotland.

M. MURRAY, R. BUTLIN, J. HOUSTON
The Building Research Establishment,
Garston, Watford. UK.

Since its initiation in the early 1970's, stone cleaning by laser ablation has gradually attracted interest as a 'self limiting' cleaning method. The late 1980's and early 1990's saw the realisation of economically viable field portable cleaning systems to be used by professionals to clean stone sculptures and decorated facades *in situ.* The success of laser cleaning limestone and marble sculpture is widely acknowledged, however the suitability of the technique for use on stone of other types requires investigation. This work describes some results from a study being undertaken to examine the potential for the laser cleaning of Scottish sandstone and granite. Bleaching thresholds for fresh cut stone exposed to pulsed radiation at 1064nm from an Nd:YAG laser were determined by change in red colouration determined by using a colour meter. From these preliminary investigations it appears that certain granites and sandstones, in particular those containing iron oxides, can change colour although the mechanisms by which colour change occurs requires elucidation. The colour change occurs at laser energy densities between 100 and 300 $mJcm^{-2}$, depending on stone type. The energy densities employed cleaning typically range from 0.1 to 10 J cm^{-2}. Those which bring about colour loss in red stone types are clearly well within the ranges used to remove black urban soiling.

Stone Weathering and Atmospheric Pollution Network '97: Aspects of Stone Weathering, Decay and Conservation.
Edited by M.S. Jones & R.D. Wakefield © 1998 Imperial College Press.

NEW INSTRUMENTS FOR MONITORING ALGAE POPULATIONS ON STONE SURFACES

E. BRECHET, D. McSTAY, R.D. WAKEFIELD, M.S. JONES
Optoelectronics Research Group, The Robert Gordon University,
School of Applied Sciences, Schoolhill, Aberdeen.
Tel: ++(0)1224-262000; Fax: ++(0)1224-162828
E mail e.brechet@rgu.ac.uk

The appearance of growth at the surface of stone samples is currently largely monitored using visual techniques. To shorten the length of time under which laboratory tests need to run and samples remain in incubation, sensitive and quantitative *in situ* on-site detection methods for measuring the initial colonising organisms are required. Some success has been achieved using colour meters, where the degree of 'greening' and 'darkening' of the surface due to microbial growth is assessed using the colour components L* a* b*.

Two novel non-destructive methods for determining algal population*in-situ* at a stone surface using fluorescence at 685nm due to chlorophyll-a are now available, one a laboratory based device for large stone samples, and a second for use on site.

The laboratory based system comprises a scanning fluorescence microscope with an Argon-ion laser as the excitation source designed to measure the 685nm fluorescence from biological growth on stone. The application of the system to the monitoring of algal growth and activity on stone by producing periodic fluorescence maps of selected test stone samples is now possible. The distribution of the measured fluorescence is shown to correlate well with the known algal distribution at the sample surfaces.

The field system comprises a hand-held fluorometer developed to detect algal growth on stone surfaces on site. The excitation source of the system is an Ultra-bright blue LED, used to induce chlorophyll-a fluorescence. Due to it's modulated excitation light, this system can be used on-site in various daylight conditions. Different concentrations of algal cultures embedded in sand were used to show that the measured 685nm fluorescence increased linearly with the quantity of algae in test samples.

Both systems have been used in laboratory and in on-site trials to determine the effect of biocide activity on chlorophyll-a fluorescence, and hence cell activity. They have proved to produce real time rapid and quantitative data about algal distribution, biocide effectiveness and cell re-colonisation following cleaning/biocide treatments.